Pharmaceutical Properties of Plants from the Genus *Securidaca* L.

This book presents the taxonomy, ethnobotany, pharmacological properties, and therapeutic potential of members of the genus *Securidaca*, particularly *Securidaca longepedunculata*. The introductory chapters highlight its significance in traditional medicine and contemporary scientific research. The book discusses the cultural and historical uses of plants from the genus *Securidaca* L. with an emphasis on *S. longepedunculata* and *S. inappedunculata*, the two most widely used medicinal plants from the genus, while examining the role in traditional medicine across various indigenous communities. It also provides a comprehensive overview of the pharmacological activities of the species, including their therapeutic potential, with dedicated sections on its antimalaria, anticancer, and dermatological properties. Furthermore, the book examines the antibacterial activities of *S. longepedunculata*, exploring its effectiveness against different bacterial strains using data from various *in vitro* studies coupled with statistical simulations and potential applications in modern medicine. It delves into the chemical constituents responsible for the medicinal properties of the plants and reviews studies on its toxicity, providing insights into safe usage and potential side effects. The book concludes by discussing the conservation status of *S. longepedunculata* (which is locally threatened and protected in some areas) and strategies for sustainable harvesting, ensuring its continued availability for future generations. This book is intended for researchers, professionals, and students in the fields of botany, ethnobotany, phytochemistry, pharmacognosy, conservation biology, ethnopharmacology, and pharmaceutical sciences.

Pharmaceutical Properties of Plants from the Genus *Securidaca* L.

Edited by
Douglas Ongeri Ochora
Anywar Godwin U

CRC Press
Taylor & Francis Group
Boca Raton London New York

CRC Press is an imprint of the
Taylor & Francis Group, an **informa** business

Designed cover image: Anywar Godwin U

First edition published 2026
by CRC Press
2385 NW Executive Center Drive, Suite 320, Boca Raton, FL 33431

and by CRC Press
4 Park Square, Milton Park, Abingdon, Oxon, OX14 4RN

CRC Press is an imprint of Taylor & Francis Group, LLC

ISBN: 9781032892900 (hbk)
ISBN: 9781032892917 (pbk)
ISBN: 9781003544463 (ebk)

DOI: 10.1201/9781003544463

Typeset in Sabon
by Newgen Publishing UK

Contents

Preface

The primary purpose of this book is to comprehensively document and analyse the pharmacological activities of plants belonging to the genus *Securidaca*, with a focus on their potential therapeutic applications. These species, widely used in traditional medicine across Africa, Asia, and South America, have demonstrated a broad range of bioactivities, including antimalarial, anti-inflammatory, antimicrobial, antioxidant, anticancer, antidiabetic, and dermatological effects. By compiling findings from ethnopharmacological studies, phytochemical analyses, and experimental pharmacology, the book aims to bridge the gap between traditional knowledge and modern drug discovery efforts.

Furthermore, the book seeks to highlight the bioactive compounds isolated from *Securidaca* species, such as xanthones, saponins, flavonoids, and alkaloids, emphasizing their potential as drug leads. It also underscores the importance of conserving medicinal plant biodiversity and encourages further research into the safety, efficacy, and standardization of *Securidaca*-based therapies. Ultimately, the book serves as a resource for pharmacologists, ethnobotanists, phytochemists, and researchers interested in the development of plant-based pharmaceuticals.

Acknowledgements

The editors/authors are grateful to "the Lord, for he is good, for his steadfast love endures forever!" (Psalms 107:1).

We acknowledge the contributions of fellow researchers, ethnobotanists, and traditional medical practitioners whose invaluable knowledge and studies on *Securidaca* species have laid the foundation for this work. Special thanks are due to our academic mentors and collaborators for their critical insights and continued encouragement throughout the development of this book. We appreciate the support of our families, especially our wives and children, for their support and encouragement towards this book's publication.

Editors

Douglas Ongeri Ochora holds a PhD in Botany from Makerere University, Uganda, an MSc in Botany (Plant Physiology and Biochemistry), and a BSc in Biology (Botany and Zoology), both from the University of Nairobi, Kenya. Douglas serves as a Lecturer in Botany in the Department of Biological Sciences, Kisii University, Kenya. He is also a postdoctoral fellow at the Preclinical Drug Development Platform (PCDDP), North-West University, South Africa, where he is doing research in pharmaceutical sciences. He actively mentors both undergraduate and postgraduate students doing research in botany.

Anywar Godwin U holds a PhD as a CARTA/DAAD fellow from Makerere University, the Fraunhofer Institute for Cell Therapy and Immunology, and the Institute for Virology, Faculty of Medicine, University of Leipzig, Germany. Godwin is currently a lecturer in the Department of Plant Sciences, Microbiology and Biotechnology at Makerere University and is a research associate at the Bernhard Nocht Institute for Tropical Medicine (BNFT), Hamburg. He is a fellow of the Global Young Academy, African Studies Association, Africa Science Leadership Programme, and the Uganda National Academy of Sciences. He is a member of several professional bodies and a winner of several international and local awards. He is the vice president of the Natural Products Network for Eastern and Central Africa, Uganda Chapter (NAPRECA) and the Uganda Textbook Academic and Non-Fiction Authors Association (UTANA). He has authored more than 100 research papers in international peer-reviewed journals and is an editor.

Chapter 1

Antimalarial and anticancer activities of plants from the genus *Securidaca* L. (Polygalaceae)

Douglas Ongeri Ochora[1*] *and Anywar Godwin U*[2]
[1]Department of Biological Sciences, School of Pure and Applied Sciences, Kisii University, Kisii, Kenya
[2]Department of Plant Sciences, Microbiology and Biotechnology, College of Natural Sciences, Makerere University, Kampala, Uganda
Ethnopharmacology and Zoopharmacognosy, Bernhard Nocht Institute for Tropical Medicine, Hamburg, Germany

1.1 INTRODUCTION

Malaria is a parasitic disease caused by six *Plasmodium* species: *Plasmodium falciparum, P. malariae, P. knowlesi, P. vivax, P. ovalecurtisi,* and *P. ovalewallikeri* (Chen et al., 2022). Of these human malaria-causing parasites, *P. falciparum* is the most lethal and therefore the main cause of the reported malaria mortalities (Song et al., 2022). The parasite is transmitted by female anopheles mosquitoes during a blood meal (Chen et al., 2022).

Malaria remains a global health burden. In 2023, there were 263 million clinical cases and 597,000 deaths due to malaria globally. This translates to about 1,636 malaria deaths daily in that year, with pregnant women and children under the age of 5 years most affected. Africa bears the brunt of the disease, accounting for 95% of malaria deaths and 94% of malaria cases worldwide (WHO-World Malaria Report, 2024). This is mainly because the African region is challenged with poverty and limited access to conventional treatment in hospitals, coupled with favorable climatic conditions that have contributed to malaria transmission in high malaria-burden regions, especially in sub-Saharan Africa (Andagalu et al., 2022).

Furthermore, the malaria situation is aggravated by the continued reports of an ever-increasing occurrence of drug-resistant malaria-causing parasites (Ouji et al., 2018). To counteract resistance and increase therapeutic efficacy, the World Health Organization (WHO) recommends the use of combination therapy. This has led to the use of artemisinin-based combination therapies (ACTs) as the standard treatment for malaria (WHO-World Malaria Report, 2024). In Africa, artemether-lumefantrine (AL), dihydroartemisinin-piperaquine (DHA-PPQ), and artesunate-amodiaquine (AS-AQ) are the recommended ACTs as the first-line treatment of uncomplicated malaria.

***Correspondence:** ochoraongeri@gmail.com

DOI: 10.1201/9781003544463-1

However, AL is the widely used ACT in most African countries (World Health Organization-Global Malaria Programme, 2019). Nevertheless, epidemiological reports have shown that *P. falciparum* has developed resistance to the currently used ACTs in Southeast Asia (Ounjaijean & Somsak, 2023) and various African countries (Conrad et al., 2023; Conrad & Rosenthal, 2019). This has exacerbated the malaria situation, especially in Africa.

Similar to malaria, cancer is the second leading cause of death globally, and the global cancer burden is expected to increase over the next two decades (Kocarnik et al., 2022). Studies have shown a potential association between communicable diseases (CDs) and non-communicable diseases (NCDs). Correspondingly, studies have reported a potential cancer–malaria connection (Ellis et al., 2021; Nordor et al., 2018). Research in developed countries has focused on cancer, whereas research in developing countries has focused on malaria (Nordor et al., 2018). A study by Ellis suggests a malaria–cancer relationship. Increased resistance to drugs and limited access to quality healthcare in both diseases have led to high incidence rates, therefore, both accounting to high global morbidity and mortality, especially in low- and middle-income countries (Ellis et al., 2021).

The malaria–cancer association can occur through the transmission of carcinogenic agents by mosquitoes when transmitting *Plasmodium* parasites during a blood meal. This occurs especially in avian and simian malaria caused by *P. knowlesi* when unnoticed carcinogenic agents are transmitted by mosquitoes (Ward & Benelli, 2016). Repeated malaria infection in childhood predisposes people to lymphoid infection in adulthood (Wyss et al., 2020). Furthermore, studies have linked Burkitt lymphoma (a childhood cancer) to falciparum malaria but not non-falciparum malaria. This type of cancer is tenfold higher in areas of stable falciparum malaria (Quintana et al., 2020). A suppressed immunity among malaria-infected individuals also predisposes them to cancer infection (Lehrer, 2010). Therefore, tackling the association between CDs and NCDs is likely to contribute to better health (Ellis et al., 2021).

Cancer is characterized by the uncontrolled growth and spread of abnormal cells, which leads to the formation of tumors that can cause death if not controlled. There are different types of cancers based on the site of infection (Cancer Research UK, 2023). Although the main cause of cancer remains unclear, the disease is associated with risk factors such as poor lifestyle, inherited genetic mutations, tobacco use (nicotine), and immune-compromised conditions. These factors may act concurrently or in a sequence, which may lead to the initiation and/or promotion of the growth of cancer cells (Siegel et al., 2019).

According to the Global Cancer Observatory, there were 19,976,499 new cancer cases and 9,743,832 deaths globally in 2022. Lung cancer and breast cancer were the most prevalent types of cancer, contributing 12.4% and 11.5%, respectively (WHO International Agency for Cancer Research,

2024). The cancer situation is worrying, with about 29.4 million new cancer cases (WHO, 2020) and 17 million deaths annually projected by the year 2030 (Jemal et al., 2009). The high cancer incidences are mainly attributed to limited anticancer drugs due to the emergence of cancers that display multidrug resistance, since chemotherapy is the main cancer treatment and management strategy in orthodox hospital medicine (Kabel & Elmaaboud, 2020). Furthermore, anticancer drug discovery and development have challenges of time, cost, and safety issues (Assaraf et al., 2019). This calls for alternative cancer treatment options.

Globally, cancer and malaria are among the leading causes of death (Ellis et al., 2021). This calls for novel antimalarial and anticancer drugs with different modes of action to counteract the effects of these diseases. In some regions of the world, such as Africa, most people are exposed to cancer and malaria with limited quality medical care. Therefore, most African communities have resorted to the use of antimalarial and anticancer plants in traditional medicine. Plants used in traditional medicine remain a potential source of antimalarial and anticancer drugs since they are likely to have unique phytochemical compounds and possibly different modes of action (Kaur & Kaur, 2017; Raimi et al., 2020).

Some of the currently used antimalarial and anticancer drugs have been derived from plants. Modern medicines such as vinflunine, vinblastine, and vincristine (anticancer), artemisinin and quinine (antimalarial drugs), and other nutraceuticals, folk medicines, pharmaceutical intermediates, and entities for synthetic drugs have been derived from plants that have been used in herbal medicine. This is because these plants contain many medicinal metabolites such as alkaloids, terpenoids, and flavonoids (Lemma et al., 2017). Furthermore, the perception is that herbal medicines are relatively safer for humans and environmentally friendly. On this basis, researchers have focused on plants in search of alternative anticancer and antimalarial drugs (G. U. Anywar et al., 2022; Kaur & Kaur, 2017; Lemma et al., 2017).

Some plants are used for the treatment of both cancer and malaria in herbal medicine. For example, in the genus *Securidaca, S. longepedunculata* is used in various African countries for the treatment of both malaria and cancer (Mongalo et al., 2015). Therefore, this chapter draws on the reported antimalarial and anticancer activities of plants from the genus *Securidaca.* Plants such as *S. longepedunculata* from Africa and *S. inappendiculata* from Asia are the most widely studied plants for their antimalarial and anticancer properties.

1.2 METHODS

A review of scientific and relevant literature was conducted on Google Scholar and the PubMed databases. The following subject headings were used as keywords: “Anticancer plants”, “antimalarial plants”, “cancer *Securidaca*”, and “malaria *Securidaca*”. Articles and abstracts written in languages other than English were excluded. Affiliation and year of publication were not considered for exclusion or inclusion.

1.3 DISCUSSION

1.3.1 Herbal remedies for malaria and cancer

Nature remains a rich reservoir of compounds for potential use in drug development. These include novel chemotypes, pharmacophores, and scaffolds for amplification into drugs for the treatment of various diseases (Dehelean et al., 2021; Veeresham, 2012). Herbal medication has remained a natural pharmacy to mankind. Various medicinal plants continue to be used in traditional medicine to treat various diseases and infections, especially in Africa and Asia. About 80% of the population in developing countries rely on plants as their primary healthcare, based on their time-tested safety and efficacy (Dehelean et al., 2021; Oyebode et al., 2016; Veeresham, 2012).

Poverty and limited accessibility to modern healthcare have led patients to rely on these plants as their first line for treatment of malaria and cancer. Various communities have therefore identified and used various plant species for traditionally treating malaria, especially in Africa and Southeast Asia, where malaria is most prevalent (Kaur & Kaur, 2017).

Sub-Saharan Africa is the most affected region with malaria globally, accounting for about 90% of the annually reported malaria cases and deaths (WHO-World Malaria Report, 2024). Traditionally, people in the region have relied on their natural environment to provide them with remedies for their ailments and diseases (Anywar et al., 2020). The use of antimalarial plants in African traditional medicine is among the most preferred means of treatment due to their accessibility, low cost, and lack of side effects (Ocan et al., 2023).

Cancer is considered as the third major cause of death in sub-Saharan Africa (Gaobotse et al., 2023). The mortality-to incidence rate ratio (MIR) is an inverse relationship with survival that has been used to make cancer comparisons on a global scale in Africa due to limited follow-up data in the region. Africa has an estimated MIR of 0.67 compared to the average global cancer MIR of 0.47 (Lawa et al., 2024). Limited access to early detection, diagnostics, treatment, and high costs for the treatment of cancer through chemotherapy have contributed to the increased reported cancer cases and deaths, especially in Africa. This has made the disease to be considered a death sentence in Africa, although it contributes only 5.9% of new global cases and 7.8% of cancer deaths (Kasahun et al., 2020; Lawa et al., 2024).

As a means to counteract challenges associated with cancer treatment, such as side effects and drug resistance, researchers have focused their search for anticancer remedies on plants used in traditional medicine as a potential source of anticancer drug development (Seca & Pinto, 2018). *Prunus africana* has been reported as the most widely used plant for the treatment of cancer, especially prostate cancer, in traditional medicine in East Africa (Anywar et al., 2020; Kudamba et al., 2024). Similarly, various plants in East Africa have been reported to be used for the treatment of breast cancer. These plants are preferred as complementary treatments for breast cancer due to their cultural appropriateness, accessibility, affordability, and sustainability (Alum et al., 2025).

1.3.2 Antimalarial and anticancer drugs derived from plants

While herbal drug preparations have been used for thousands of years, the application of isolated and characterized compounds in modern drug discovery and development started only in the 19th century (Veeresham, 2012). These biologically active constituents act as a lead to the discovery of modern medicine (Seca & Pinto, 2018; Yuan et al., 2016). This is because secondary metabolites from plants have different structural arrangements and properties, which have helped researchers to derive synthetic or semi-synthetic medicines from plants. Studies have shown that more than 50% of modern medicines are directly or indirectly derived from natural products. However, there is a need for continued research on antimalarial and anticancer plants used in traditional medicine (Okoro et al., 2010). Researchers have focused on herbal medicine as a potential source of antimalarial and anticancer compounds (Ocan et al., 2023). This has led to the success of some of the currently used antimalarial (Table 1.1) and anticancer drugs (Table 1.2).

1.3.3 Ethnobotany of *S. longepedunculata* for the treatment of malaria

Plants belonging to the genus *Securidaca* are among the most preferred plants for the treatment of malaria in Africa. The two plant species in the genus that occur in Africa are *S. longepedunculata* and *S. welwitschii*. *S. longepedunculata* is the most preferred plant for treating malaria in traditional medicine in various African countries (Table 1.1), where it is considered a common antimalarial plant (Chinsembu, 2015).

In Kenya, a decoction prepared from the roots of *S. longepedunculata* is taken orally three times daily for 3–4 days for the treatment of malaria (Nguta et al., 2010). A combination of dry powdered roots of *S. longepedunculata* and *Zanthoxylum zanthoxyloides* (Lam.) Zepern. & Timler is sniffed or burnt, and the resulting smoke is inhaled to treat malaria and fever in traditional medicine among the Baganda people of Uganda (Hamill et al., 2003; Mongalo et al., 2015). Similarly, *S. longepedunculata* is used for the treatment of fevers and malaria (Schultz et al., 2021). A decoction prepared

Table 1.1 Antimalarial drugs derived from plants

Antimalarial Drugs	*Plant Source*	*Use*	*References*
1. Quinine	*Cinchona* L.	Treatment of severe malaria	Tse et al., 2019; WHO, 2023
2. Chloroquine	Quinine derivative	Treatment of non-falciparum malaria (*P. vivax* malaria)	
3. Artemisinin	*Artemisia annua* L.	NA	
4. Artemether	*A. annua* Artemisinin derivative	Used in combination with lumefantrine for the treatment of uncomplicated falciparum malaria	
5. Artesunate		Used in combination with mefloquine for the treatment of uncomplicated falciparum malaria	
6. Arteether		Treatment of falciparum and cerebral malaria	
7. Dihydroartemisinin		Used in combination with piperaquine for the treatment of uncomplicated falciparum malaria	

from the roots of *S. longepedunculata* is thepreferred therapy for the treatment of malaria (Table 1.3).

1.3.4 Ethnobotany of *S. longepedunculata* and *S. inappendiculata* for the treatment of cancer

Plants belonging to the genus *Securidaca* have been used for the treatment of cancer in traditional medicine. The most widely used anticancer plants of the genus include *S. longepedunculata* and *S. inappendiculata*. In Nigeria, the root bark of *S. longepedunculata* has been used for the treatment of various types of cancer in traditional medicine (Gaobotse et al., 2023; Ngulde et al., 2019), while fresh leaves have been used for the treatment of skin cancer (Mustapha, 2013) and pancreatic cancer (Raimi et al., 2020) and root decoction for breast cancer (Ashidi et al., 2010). Furthermore, *S. longepedunculata* has been used in traditional medicine for the treatment of cancer in Togo (Kola et al., 2023) and Congo (Dibwe et al., 2014). Similarly, *S. inappendiculata* has been widely used in Chinese traditional medicine for the treatment of various types of cancer (lan Ji et al., 2021; Zha et al., 2015). These suggest the potential use of these plants in the treatment of various types of cancer.

Table 1.2 Anticancer drugs derived from plants

Anticancer Drugs	*Plant Source*	*Use*	*References*
1. Vincristine	*Catharanthus roseus* (L.) G. Don	Leukaemia, Lymphoma, Hodgkin's disease, Erythremia, Panmyelosis, breast cancer, lung cancer	Haque et al., 2016; Moudi et al., 2013; Ojima et al., 2016; DrugBank, 2025
2. Vinblastine		Breast cancer, testicular cancer, lymphoma, and Kaposi's sarcoma	
3. Vinorelbine	*C. roseus* Derivative of vinblastine	Lung cancer Lymphoma, head and neck cancer, ovarian cancer, breast cancer, osteosarcoma, lung cancer	
4. Vinflunine	*C. roseus* Derivative of vincristine	Lung cancer, breast cancer	
5. Paclitaxel	*Taxus brevifolia* Nutt.	Kaposi's sarcoma and cancer of the lung, ovarian, and breast	
6. Docetaxel	*Taxus baccata* L. Derivative of paclitaxel	Breast cancer, lung cancer, prostate cancer, carcinoma of the head and neck	
7. Cabazitaxel		Prostate cancer	
8. Camptothecin	*Camptotheca acuminata* Decne.	Colorectal, ovarian, and lung cancer	
9. Topotecan		Lung cancer, cervical cancer, pediatric cancer	
10. Irinotecan		Colon or rectum cancer	
11. Etoposide	*Podophyllum peltatum* L.	Lung cancer	

1.3.5 Antimalarial activities of *S. longepedunculata*

In the genus *Securidaca*, only *S. longepedunculata* has been studied for its antimalarial activities. This is mainly because it occurs in Africa, where malaria is most prevalent, and is used for malaria treatment in traditional herbal medicine (Table 1.3). In a recent pre-screen assay of water, methanol, and ethyl acetate stem extracts of *S. longepedunculata*, all tested at 1,000 μg/mL, all the extracts did not show hemozoin formation inhibition activity in a heme biocrystallization library screening against *P. falciparum* K1 strain (Schultz et al., 2021). Further *ex vivo, in vitro,* and *in vivo* antiplasmodial assays of the plant have shown its potential use as an antimalarial drug, especially the roots. *In vitro* and *ex vivo* antiplasmodial activities for crude

Table 1.3 Use of *S. longepedunculata* for the treatment of malaria in Africa

Part of the Plant	*Mode of Use*	*Country Used*	*References*
1. Roots	Decoction	Kenya	Nguta et al., 2010; Ochora et al., 2022
		Uganda	Hamill et al., 2003; Opio et al., 2018; Schultz et al., 2021
		Tanzania	Joseph et al., 2006
		Mali and São Tomé	Ancolio et al., 2002; Diarra et al., 2015; Mukungu et al., 2016
		Angola	Novotna et al., 2020
		Malawi	Nguta et al., 2010; Williamson, 1972
		Zambia	Chinsembu, 2015
		South Africa	
		Namibia	
2. Root bark	Decoction	Nigeria	Mustapha, 2013; Chinsembu, 2015
3. Roots and stem bark		Burkina Faso	Nadembega et al., 2011
4. Leaves	Decoction	Mali	Diarra et al., 2015
	NA	Ghana	Ohashi et al., 2018

NA: Not available.

plant extracts and pure compounds can be discussed in terms of the 50% inhibitory concentration (IC_{50}) values, as described in Table 1.4.

One of the earlier studies on antimalarial activities of *S. longepedunculata* by Bah et al. (2007) reported good antiplasmodial activities of dichloromethane leaf extracts of the plant, with an IC_{50} value of 7 µg/mL when tested against *P. falciparum* (3D7). In the same study, water leaf extracts of *S. longepedunculata* showed activity of more than 40% parasite growth inhibition when tested against *P. falciparum* (3D7) (Bah et al., 2007). A similar study reported a low *in vitro* antiplasmodial activity of ethyl alcohol leaf extracts of the plant with an IC_{50} value of 114.4 µg/mL when tested against the same *P. falciparum* strain, 3D7 (Ohashi et al., 2018).

In our previous studies, both the ethyl acetate (EtOAc) and methanol (MeOH) extracts from leaves and roots of *S. longepedunculata* were tested for antiplasmodial activities using both clinical and cultured strains of *P. falciparum* (Ochora et al., 2021). The EtOAc and MeOH root extracts displayed higher antiplasmodial activity than the leaf extracts of the plant species. This validates the widespread use of the roots for the treatment of malaria in African traditional herbal medicine. The MeOH extracts of the roots showed potent antiplasmodial activities against the W2 (IC_{50} 1.4 ng/mL), D6 (IC_{50} 2.9 ng/mL), and 3D7 (IC_{50} 3.6 ng/mL) cultured strains of

Table 1.4 Levels of *in vivo* and *ex vivo* antiplasmodial activities in IC_{50} for crude *Securidaca longepedunculata* plant extracts and pure compounds

Range	*Activity of Crude Extracts (μg/mL)*	*References*
1. > 100	Inactive	Basco et al., 1994
2. 50–100	Low activity	
3. 10–50	Moderately activity	
4. <10	Good activity	
	Activity of Pure Compounds (μM)	
5. > 200	Inactive	Batista et al., 2009; Namukobe et al., 2015
6. 200–100	Low activity	
7. 20–100	Moderately activity	
8. 1–20	Good activity	
9. <1	Potent/excellent	

P. falciparum. Similarly, the EtOAc extract of the roots also displayed good antiplasmodial activities, with IC_{50} values of < 5 ng/mL against the D6 and 3D7 cultured strains of *P. falciparum* and an IC_{50} value of 9.8 ng/mL when tested *ex vivo* using clinical isolates (tested with 6 hours of blood collection from malaria patients). The leaf extracts exhibited moderate antiplasmodial activity across clinical isolates and cultured *P. falciparum* strains (Ochora et al., 2021). This indicates a potential use of the root extracts of the plant as an additional source to explore in antimalarial drug discovery.

Likewise, pure compounds isolated from the roots of *S. longepedunculata* were characterized and tested for antiplasmodial activities. The eight pure compounds isolated from the roots exhibited good to moderate antiplasmodial activities against clinical isolates and cultured strains of *P. falciparum,* with IC_{50} values of 10–100 μM (Ochora et al., 2021). We showed that 4-hydroxy-2,3-dimethoxybenzophenone was the most active compound with an IC_{50} value of 18.6 μM when tested *ex vivo* against field isolates. Extracts exhibited higher antiplasmodial activities than pure compounds (Ochora et al., 2021). This can be attributed to the synergism of various compounds in the extracts. This is the only study that reported the antiplasmodial activities of pure compounds of *S. longepedunculata.* Considering the use of the plant in the treatment of malaria in various countries, antiplasmodial activities of other compounds from the genus *Securidaca* should be explored for their potential use in antimalarial drug discovery and development.

In another follow-up study, the MeOH leaves, stems, and root extracts of *S. longeipedunculata* were each combined with artemether (ART) and lumefantrine (LU) and tested against cultured strains of *P. falciparum* (D6,

W2, and DD2) and clinical isolates. Synergism was reported in most combinations at all fixed extract–drug ratios of 4:1, 3:1, 1:1, 1:2, 1:3, and 1:4. Antiplasmodial activities of combinations are reported in mean sum 50% fractional inhibitory concentration (FIC_{50}). The FIC_{50} can be displayed as synergism, ($FIC_{50} <1$), where one drug potentiates the activity of another drug resulting in higher activity than using one drug alone, additivity ($FIC_{50} = 1$) where the combination of the two drugs results in the same activity as when the individual drugs are used alone, and antagonism ($FIC_{50} > 1$) where the activity of the two drugs in combination is lower than that of the individual drugs when used alone (Ochora et al., 2023; Ochora et al., 2022).

A combination of the MeOH root extracts with ART showed synergism, with a mean of mean FIC_{50} of 0.4 across all fixed extract–drug ratios when tested against D6 strain of *P. falciparum*. Correspondingly, when leaf extracts were combined with LU and tested against the W2 strain, synergism was observed with a mean of mean FIC_{50} of 0.46. The MeOH stem extracts combined with LU against the DD2 strain displayed synergism, with a mean of mean FIC_{50} of 0.38. Furthermore, all fixed ratios of stem extracts with ART showed synergism when tested *ex vivo* against fresh field isolates (Ochora et al., 2022). These results show that the use of antimalarial plants such as *S. longepedunculata* in combination with antimalarial drugs is likely to increase the overall antimalarial efficacy. This reflects the herb–drug interactions that are likely to be caused by the concurrent use of antimalarial plants and drugs, especially in Africa (Erhirhie et al., 2021).

Methanol root extracts of *S. longepedunculata* were tested against *P. berghei* (NK65) using Swiss albino mice. The extract displayed chemosuppression activities of 82.6%, 62.7%, and 58.4% at concentrations of 0.56 mg/kg, 0.28 mg/kg, and 0.14 mg/kg, respectively (Haruna et al., 2013). In a similar study, antiplasmodial activities of water and organic (chloroform and methanol at 1:1, v/v) root extracts of *S. longepedunculata* against *P. berghei* (ANKA) were determined using Swiss albino mice. The organic root extracts displayed a high parasite chemosuppression activity of 91.03%, and the water extracts showed a chemosuppression activity of 55.05% (Ochora et al., 2014). The observed *in vivo* antiplasmodial activities confirm similar *in vitro* studies, which suggest the possible use of the plant as a source of antimalarial compounds that can act as a scaffold for antimalarial drug discovery and development.

Only one study has reported the *in vitro* and *ex vivo* antiplasmodial activities of pure compounds from the roots of *S. longepedunculata* (Ochora et al., 2021). Most of the previous studies on *Securidaca* spp. isolated and characterized pure compounds from the plants without carrying out bioassays to determine the pharmaceutical properties of the isolated compounds (Lognay et al., 2000; Tommasi et al., 1993; Zhang et al., 2005; Kang et al., 2005; Kang et al., 2009; Yang et al., 2002). However, most of the current research has focused on the medicinal properties of compounds isolated from *Securidaca* spp., especially their anticancer properties.

1.3.6 Anticancer activities of plants from the genus *Securidaca*

The limitations of conventional chemotherapy, such as adverse side effects and the ever-increasing occurrence of multidrug resistance against anticancer drugs, have increased the focus of the field of oncology on anticancer plants (Dehelean et al., 2021). This has led cancer patients to resort to herbal medicine for the treatment of cancer, especially in Africa (Adico et al., 2024; Gaobotse et al., 2023). There are a large number of phytochemicals from these anticancer plants in Africa undergoing preclinical and clinical studies as potential anticancer agents (Adico et al., 2024; Seca & Pinto, 2018). Among the most widely studied anticancer plants in the genus *Securidaca* are *S. longepedunculata* and *S. inappendiculata*.

1.3.7 Anticancer activities of *S. longepedunculata* and *S. inappendiculata*

Various pure compounds isolated from *S. longepedunculata*, especially the roots, have been characterized and tested for their anticancer activities. Xanthones, benzyl benzoates, bisdesmosidic saponins, methyl salicylate, and triterpene saponins from plants have shown antitumor and cytotoxic activities (Mitaine-offer et al., 2010; Obasi et al., 2020; Zuo et al., 2014).

In a study by Zha et al. (2015), securioside C and securioside D from *S. longepedunculata* displayed moderate *in vitro* cytotoxic activity against Lewis lung carcinoma cells with IC_{50} values of 45.56 μM and 85.98 μM, respectively (Zha et al., 2015). Similarly, two compounds, 1,6,8-trihydroxy-2,3,4,5-tetramethoxyxanthone and 1,6-dihydroxy-2,3,4,5,8-pentamethoxyxanthone, from the roots of *S. longepedunculata,* exhibited *in vitro* preferential cytotoxicity, with PC_{50} values of 22.8 μM and 17.4 μM, respectively, against the human pancreatic cancer cell line, PANC-1 (Dibwe et al., 2013). Furthermore, saponins from *S. longepedunculata* showed induced apoptosis and inhibited the spread and invasion of cervical cancer cells (Obasi et al., 2020). In our previous study, benzophenones, 2-hydroxy-6-methoxybenzoic acid, and 2,4,5-trimethoxybenzophenone, isolated from root extracts of *S. longepedunculata,* displayed moderate *in vitro* anticancer activities against human lung epithelial A549 cells (IC_{50} =22.2 μM) and liver cancer cells HepG2 (IC_{50} 45.2=μM), respectively (Ochora et al., 2021).

These studies continue to report moderate to potent anticancer activities of compounds isolated from *S. longepedunculata* across different cell lines. However, it is not obvious that compounds from *S. longepedunculata* have anticancer activity, as some compounds have been reported with no anticancer activity. For example, substituted xanthones from the roots of *S. longepedunculata* displayed no cytotoxic activity when tested *in vitro* against both human pancreatic cancer PANC-1 cells in nutrient-rich and nutrient-deprived media (Dibwe et al., 2014). This confirms the necessity of continued screening of compounds from plants to ascertain their acclaimed therapeutic activities.

Further studies on extracts of *S. longepedunculata* have been conducted for their anticancer activities. The methanol root extracts of the plant species showed moderate *in vitro* cytotoxic activities against A549 lung (IC_{50} 66.4 µg/mL) and HepG2 liver (IC_{50} 77.5 µg/mL) cancerous cell lines (Ochora et al., 2021). Ethanolic extracts from the root bark of *S. longepedunculata* showed potential activity to repress the proliferation of the U87 brain tumor cell line, with an IC_{50} value of 20.54 µg/mL. The extracts induced apoptosis at 41.53% poly-ADP-ribose-polymerase cleavage (Ngulde et al., 2019). These *in vitro* anticancer activities for both extracts and pure compounds should be further explored in animal models, as other physiological factors and diet may influence the observed therapeutic activities.

Water root back extracts of *S. longepedunculata* exhibited *in vitro* and *in vivo* cytotoxic activities using Swiss albino mice (Lawal et al., 2012). Similarly, hydroethanolic extracts of *S. longepedunculata* showed high *in vitro* and *in vivo* anti-inflammatory activity comparable to the standard drugs (staurosporine and doxorubicin). When tested *in vitro*, the extracts exhibited anticancer activity with an IC_{50} value of 150 µg/mL when tested against lung cancer cell lines. The study further displayed a reduced tumor volume of 2.84 mL and 2.77 mL when tested at 80 mg/kg and 160 mg/kg, respectively, against murine cancer cell lines Enrlich Ascites carcinoma using BALB/c mice models. This resulted in an increased survival and life expectancy of the experimental animals (Kola et al., 2023). These preclinical studies indicate the potential use of the secondary metabolites from plants in cancer treatment (Seca & Pinto, 2018).

Likewise, various studies have reported anticancer activities of compounds isolated from *S. inappendiculata*. Xanthones isolated from stems of *S. inappendiculata* showed *in vitro* activity against the human lung adenocarcinoma epithelial cell line A549. The study showed a structure-activity concentration-dependent activity, with the highest activity displayed by 1,7-dihydroxy-3,4-dimethoxyxanthone with an IC_{50} value of 50.66 µM (Zuo et al., 2014). Hemiterpene and furolactone-type lignan glycosides from roots of *S. inappendiculata* showed cytotoxic activities with IC_{50} values of 145.3 µM and 57.5 µM, respectively, when tested for anti-inflammatory activity against nitric oxide overproduction in lipopolysaccharide-stimulated mouse macrophage RAW264.7 model. Therefore, none of the compounds showed obvious cytotoxic activities (Yang et al., 2020). Similarly, three lignan glycosides showed moderate (IC_{50} = 20–100 µM) hepatoprotective *in vitro* activity when tested for acetaminophen-induced HepG2 cell injury (Ji et al., 2019). These studies emphasize the potential use of these compounds as scaffolds for anticancer drug discovery and development. Regardless of the current advances in synthetic chemistry, nature remains the main source of pharmaceutical agents (Dehelean et al., 2021). This underscores the role of phytotherapy in the treatment of malaria and cancer.

1.4 CONCLUSION

Herbal medications have contributed to the management of malaria and cancer through their continued use in traditional medicine and drugs derived from plants. This chapter reveals the antimalarial and anticancer properties of plants from the genus *Securidaca*. The most studied plants are *S. longepedunculata* and *S. inappendiculata*. Most studies have reported the anticancer and antimalarial activities of these two plant species. Pure compounds from these plants showed potential use of these plants as anticancer agents. These should further be confirmed in various animal models and clinical trials. Extracts of *S. longepedunculata* showed higher antimalarial activities than pure compounds. Antimalarial activities of *S. inappendiculata* and other plants of the genus remain unknown, yet the phytochemistry of the plants is likely to be similar to *S. longepedunculata*. Furthermore, pharmacokinetic parameters of compounds from the plants should be determined; this will act as an initial guide for dosage, especially for the treatment of malaria in children and pregnant women. Roots are the most studied parts of these plants, which indicates urgent conservation measures for sustainable utilization. We hope this chapter will be used as a source of preliminary information for further antimalarial and anticancer studies of plants from the genus *Securidaca*.

REFERENCES

Adico, M. D. W., Bayala, B., Bunay, J., Baron, S., Simpore, J., & Lobaccaro, J. M. A. (2024). Contribution of Sub-Saharan African medicinal plants to cancer research: Scientific basis 2013–2023. *Pharmacological Research*, *202*, 107138. https://doi.org/10.1016/j.phrs.2024.107138

Alum, E. U., Tufail, T., Uti, D. E., Aja, P. M., Offor, C. E., Ibiam, U. A., Ukaidi, C. U. A., & Alum, B. N. (2025). Utilizing indigenous flora in East Africa for breast cancer treatment: An overview. *Anti-Cancer Agents in Medicinal Chemistry*, *25*(2), 99–113. https://doi.org/10.2174/0118715206338557240909081833

Ancolio, C., Azas, N., Mahiou, V., Ollivier, E., Di Giorgio, C., Keita, A., Timon-David, P., & Balansard, G. (2002). Antimalarial activity of extracts and alkaloids isolated from six plants used in traditional medicine in Mali and Sao Tome. *Phytotherapy Research*, *16*(7), 646–649. https://doi.org/10.1002/ptr.1025

Andagalu, B., Watson, O. J., Onyango, I., Opot, B., Okoth, R., Chemwor, G., Sifuna, P., Juma, D., Cheruiyot, A., Yeda, R., Okudo, C., Wafubwa, J., Yalwala, S., Abuom, D., Ogutu, B., Cowden, J., Akala, H. M., & Kamau, E. (2022). Malaria transmission dynamics in a high-transmission setting of Western Kenya and the inadequate treatment response to artemether-lumefantrine in an asymptomatic population. *Clinical Infectious Diseases*, *76*(4), 704–712. https://doi.org/10.1093/cid/ciac527

Anywar, G., Kakudidi, E., Byamukama, R., Mukonzo, J., Schubert, A., & Oryem-Origa, H. (2020). Indigenous traditional knowledge of medicinal plants used by herbalists in treating opportunistic infections among people living with HIV/AIDS in Uganda. *Journal of Ethnopharmacology*, *246*, 112205. https://doi.org/10.1016/J.JEP.2019.112205

Anywar, G. U., Kakudidi, E., Oryem-Origa, H., Schubert, A., & Jassoy, C. (2022). Cytotoxicity of medicinal plant species used by traditional healers in treating people suffering from HIV/AIDS in Uganda. *Frontiers in Toxicology*, *4*(May), 1–8. https://doi.org/10.3389/ftox.2022.832780

Ashidi, J. S., Houghton, P. J., Hylands, P. J., & Efferth, T. (2010). Ethnobotanical survey and cytotoxicity testing of plants of South-western Nigeria used to treat cancer, with isolation of cytotoxic constituents from Cajanus cajan Mill sp. leaves. *Journal of Ethnopharmacology*, *128*(2), 501–512. https://doi.org/10.1016/j.jep.2010.01.009

Assaraf, Y. G., Brozovic, A., Gonçalves, A. C., Jurkovicova, D., Linē, A., Machuqueiro, M., Saponara, S., Sarmento-ribeiro, A. B., Xavier, C. P. R., & Vasconcelos, M. H. (2019). The multi-factorial nature of clinical multidrug resistance in cancer. *Drug Resistance Updates*, *46*(August), 100645. https://doi.org/10.1016/j.drup.2019.100645

Bah, S., Jäger, A. K., Adsersen, A., Diallo, D., & Paulsen, B. S. (2007). Antiplasmodial and GABAA-benzodiazepine receptor binding activities of five plants used in traditional medicine in Mali, West Africa. *Journal of Ethnopharmacology*, *110*(3), 451–457. https://doi.org/10.1016/j.jep.2006.10.019

Basco, L. K., Mitaku, S., Skaltsounis, A. L., Ravelomanantsoa, N., Tillequin, F., Koch, M., & Le Bras, J. (1994). *In vitro* activities of furoquinoline and acridone alkaloids against *Plasmodium falciparum*. *Antimicrobial Agents and Chemotherapy*, *38*(5), 1169–1171. https://doi.org/10.1128/AAC.38.5.1169

Batista, R., De Jesus Silva Júnior, A., & De Oliveira, A. B. (2009). Plant-derived antimalarial agents: New leads and efficient phytomedicines. part II. non-alkaloidal natural products. *Molecules*, *14*(8), 3037–3072. https://doi.org/10.3390/molecules14083037

Cancer Research, UK (2023). Cancer in the UK Overview 2023. https://actionkidneycancer.org/wp-content/uploads/2023/02/Cancer-in-the-UK-Overview-2023.pdf (accessed on 17 December 2025).

Chen, J., Ma, X., Tang, J., Xu, S., Gu, Y., Tang, F., Cao, Y., Wang, W., Zhou, H., Zhang, J., Yu, X., Zhu, G., Zhu, M., Zhang, Q., Culleton, R., Liu, Y., & Cao, J. (2022). Disparate selection of mutations in the dihydrofolate reductase gene (dhfr) of *Plasmodium ovale curtisi* and *P. o. wallikeri* in Africa. *PLoS Neglected Tropical Diseases*, *16*(12), 1–14. https://doi.org/10.1371/journal.pntd.0010977

Chinsembu, K. C. (2015). Plants as antimalarial agents in Sub-Saharan Africa. *Acta Tropica*, *152*, 32–48. https://doi.org/10.1016/j.actatropica.2015.08.009

Conrad, M. D., Asua, V., Garg, S., Giesbrecht, D., Niaré, K., Smith, S., Namuganga, J. F., Katairo, T., Legac, J., Crudale, R. M., Tumwebaze, P. K., Nsobya, S. L., Cooper, R. A., Kamya, M. R., Dorsey, G., Bailey, J. A., & Rosenthal, P. J. (2023). Evolution of partial resistance to artemisinins in malaria parasites in Uganda. *New England Journal of Medicine*, *389*(8), 722–732. https://doi.org/10.1056/nejmoa2211803

Conrad, M. D., & Rosenthal, P. J. (2019). Antimalarial drug resistance in Africa: The calm before the storm? *The Lancet Infectious Diseases*, *19*(10), e338–e351. https://doi.org/10.1016/S1473-3099(19)30261-0

De Tommasi, N., Piacente, S., Suione, F. D. E., & Pizza, C. (1993). New sucrose derivatives from the bark of *Securidaca longipedunculata*. *Journal of Natural Products*, *56*(1), 134–137.

Dehelean, C. A., Marcovici, I., Soica, C., Mioc, M., Coricovac, D., Iurciuc, S., Cretu, O. M., & Pinzaru, I. (2021). Drug discovery and alternative therapy. *Molecules*, *1096*, 1–29.

Diarra, N., Klooster, C. V. T., Togola, A., Diallo, D., Willcox, M., & De Jong, J. (2015). Ethnobotanical study of plants used against malaria in Sélingué subdistrict, Mali. *Journal of Ethnopharmacology*, *166*, 352–360. https://doi.org/10.1016/j.jep.2015.02.054

Dibwe, D. F., Awale, S., Kadota, S., Morita, H., & Tezuka, Y. (2013). Heptaoxygenated xanthones as anti-austerity agents from Securidaca longepedunculata. *Bioorganic and Medicinal Chemistry*, *21*(24), 7663–7668. https://doi.org/10.1016/j.bmc.2013.10.027

Dibwe, D. F., Awale, S., Kadota, S., Morita, H., & Tezuka, Y. (2014). Muchimangins G-J, fully substituted xanthones with a diphenylmethyl substituent, from *Securidaca longepedunculata*. *Journal of Natural Products*, *77*(5), 1241–1244. https://doi.org/10.1021/np5000445

DrugBank (2025). Vincristine. https://go.drugbank.com/drugs/DB00541 (accessed 17 December 2025).

Ellis, T., Eze, E., & Tolulope Raimi-Abraham, B. (2021). Malaria and Cancer: A critical review on the established associations and new perspectives. *Infectious Agents and Cancer*, *16*(33), 1–14. https://doi.org/10.1186/s13027-021-00370-7

Erhirhie, E. O., Ikegbune, C., Okeke, A. I., Onwuzuligbo, C. C., Madubuogwu, N. U., Chukwudulue, U. M., & Okonkwo, O. B. (2021). Antimalarial herbal drugs: A review of their interactions with conventional antimalarial drugs. *Clinical Phytoscience*, *7*(4), 1–10. https://doi.org/10.1186/s40816-020-00242-4

Gaobotse, G., Venkataraman, S., Brown, P. D., Masisi, K., Kwape, T. E., Nkwe, D. O., Rantong, G., & Makhzoum, A. (2023). The use of African medicinal plants in cancer management. *Frontiers in Pharmacology*, *14*, 1122388. https://doi.org/10.3389/FPHAR.2023.1122388

Hamill, F. A., Apio, S., Mubiru, N. K., Bukenya-Ziraba, R., Mosango, M., Maganyi, O. W., & Soejarto, D. D. (2003). Traditional herbal drugs of Southern Uganda, II: Literature analysis and antimicrobial assays. *Journal of Ethnopharmacology*, *84*(1), 57–78. https://doi.org/10.1016/S0378-8741(02)00289-1

Haque, M. U., Ferdiousi, N., & Sajon, S. R. (2016). Anti-cancer agents derived from plant and dietary sources: A review. *International Journal of Pharmacognosy*, *3*(2), 55–66. https://doi.org/10.13040/IJPSR.0975-8232.IJP.3(2).55-66

Haruna, Y., Kwanashie, H. O., Anuka, J. A., Atawodi, S. E., & Hussaini, I. M. (2013). *In vivo* Anti-malarial Activity of Methanol Root Extract of *Securidaca longepedunculata* in Mice Infected with *Plasmodium berghei. International Journal of Modern Biology and Medicine*, *3*(1), 7–16. www.ModernScientificPress.com/Journals/IJBioMed.aspx

Jemal, A., Siegel, R., Ward, E., Hao, Y., Xu, J., & Thun, M. J. (2009). Cancer statistics, 2009. *Cancer Journal for Clinicians*, *59*(4), 225–249. https://doi.org/10.3322/caac.20006

Ji, C. Lan, Dai, S., Liu, H., Dong, J. Yang, Liu, C. Sheng, & Zuo, J. (2021). Polyphenols from *Securidaca inappendiculata* alleviated acute lung injury in rats by inhibiting oxidative stress sensitive pathways. *Chinese Herbal Medicines*, *13*(3), 381–388. https://doi.org/10.1016/j.chmed.2020.09.007

Ji, J., Wang, Q., Wang, M., Chen, J., & Li, X. (2019). New lignan glycosides from the stems of *Securidaca inappendiculata* Hassk. *Phytochemistry Letters*, *31*(138), 58–62. https://doi.org/10.1016/j.phytol.2019.03.011

Joseph, C. C., Moshi, M. J., Sempombe, J., & Nkunya, M. H. H. (2006). (4-methoxy-benzo[1,3]dioxol-5-yl)-phenylmethanone: An antibacterial benzophenone from *Securidaca longepedunculata. African Journal of Traditional, Complementary and Alternative Medicines, 3*(3), 80–86.

Kabel, A. M., & Elmaaboud, M. A. A. (2020). Resistance to Anticancer Agents: Recent Trends. *Journal of Cancer Research and Treatment*, *8*(2), 10–14. https://doi.org/10.12691/jcrt-8-2-1

Kang, W., Wang, Z., Li, Z., & Xu, X. (2005). Three new compounds from *Securidaca inappendiculata. Helvetica*, *88*, 2771–2776. https://doi.org/10.1002/hlca.200590217

Kang, W. Y., Ji, Z. Q., & Wang, J. M. (2009). A new xanthone from the roots of *Securidaca inappendiculata. Chemical Papers*, *63*(1), 102–104. https://doi.org/10.2478/s11696-008-0087-y

Kasahun, G. G., Gebretekle, G. B., Hailemichael, Y., Woldemariam, A. A., & Fenta, T. G. (2020). Catastrophic healthcare expenditure and coping strategies among patients attending cancer treatment services in Addis Ababa, Ethiopia. *BMC Public Health*, *20*(1), 1–10. https://doi.org/10.1186/s12889-020-09137-y

Kaur, R., & Kaur, H. (2017). Plant-derived antimalarial agents. *Journal of Medicinal Plants Studies*, *5*(1), 346–363. www.plantsjournal.com/archives/2017/vol5issue1/PartE/5-1-7-994.pdf

Kocarnik, J. M., Compton, K., Dean, F. E., Fu, W., Gaw, B. L., Harvey, J. D., Henrikson, H. J., Lu, D., Pennini, A., Xu, R., Ababneh, E., Abbasi-Kangevari, M., Abbastabar, H., Abd-Elsalam, S. M., Abdoli, A., Abedi, A., Abidi, H., Abolhassani, H., Adedeji, I. A., … Force, L. M. (2022). Cancer incidence, mortality, years of life lost, years lived with disability, and disability-adjusted life years for 29 cancer groups from 2010 to 2019: a systematic analysis for the global burden of disease study 2019. *JAMA Oncology*, *8*(3), 420–444. https://doi.org/10.1001/JAMAONCOL.2021.6987,

Kola, P., Manjula, S. N., Metowogo, K., Madhunapantula, S. V., & Eklu-Gadegbeku, K. (2023). Four Togolese plant species exhibiting cytotoxicity and antitumor activities lightning polytherapy approach in cancer treatment. *Heliyon*, *9*(3), e13869. https://doi.org/10.1016/J.HELIYON.2023.E13869

Kudamba, A., Kasolo, J. N., Bbosa, G. S., Lugaajju, A., Wabinga, H., Kafeero, H. M., Ssenku, J. E., Alemu, S. O., Walusansa, A., Niyonzima, N., & Muwonge, H. (2024). Review of herbal medicinal plants used in the management of cancers in the East Africa region from 2019 to 2023. *Integrative Cancer Therapies*, *23*, 15347354241235583. https://doi.org/10.1177/15347354241235583

Lawa, Q. O., Lawal, I., Eigbefoh, J., Aigbonoga, D. E., & Lawal, G. (2024). Cancer care gap: Examining cancer mortality-incidence rate ratio in sub-saharan Africa. *JCO Global Oncology*, *10*(1), 35–35. https://doi.org/10.1200/GO-24-32000

Lawal, R. A., Ozaslan, M. D., Odesanmi, O. S., Karagoz, I. D., Kilic, I. H., & Ebuehi, O. A. T. (2012). Cytotoxic and antiproliferative activity of *Securidaca longepedunculata* aqueous extract on Ehrlich ascites carcinoma cells in Swiss albino mice. *International Journal of Applied Research in Natural Products*, *5*(4), 19–27.

Lehrer, S. (2010). Association between malaria incidence and all cancer mortality in fifty U.S. states and the District of Columbia. *Anticancer Research*, *30*(4), 1371–1373.

Lemma, M. T., Ahmed, A. M., Elhady, M. T., Ngo, H. T., Vu, T. L. H., Sang, T. K., Campos-Alberto, E., Sayed, A., Mizukami, S., Na-Bangchang, K., Huy, N. T., Hirayama, K., & Karbwang, J. (2017). Medicinal plants for *in vitro* antiplasmodial activities: A systematic review of literature. *Parasitology International*, *66*(6), 713–720. https://doi.org/10.1016/j.parint.2017.09.002

Lognay, G., Marlier, M., Seck, D., & Haubruge, É. (2000). The occurrence of 2-hydroxy-6-methoxybenzoic acid methyl ester in *Securidaca longepedunculata* Fresen root bark. *Biotechnologie, Agronomie, Société et Environnement*, *4*(2), 107–110.

Mitaine-offer, A., Pénez, N., Miyamoto, T., Delaude, C., Mirjolet, J., Duchamp, O., & Lacaille-dubois, M. (2010). Acylated triterpene saponins from the roots of *Securidaca longepedunculata*. *Phytochemistry*, *71*(1), 90–94. https://doi.org/10.1016/j.phytochem.2009.09.022

Mongalo, N. I., McGaw, L. J., Finnie, J. F., & Staden, J. Van. (2015). *Securidaca longipedunculata* Fresen (Polygalaceae): A review of its ethnomedicinal uses, phytochemistry, pharmacological properties, and toxicology. *Journal of Ethnopharmacology*, *165*, 215–226. https://doi.org/10.1016/j.jep.2015.02.041

Moudi, M., Go, R., Yien, C. Y. S., & Nazre, M. (2013). Vinca alkaloids. *International Journal of Preventive Medicine*, *4*(11), 1131–1135. https://doi.org/10.2165/00128415-200711380-00080

Mukungu, N., Abuga, K., Okalebo, F., Ingwela, R., & Mwangi, J. (2016). Medicinal plants used for management of malaria among the Luhya community of Kakamega East sub-County, Kenya. *Journal of Ethnopharmacology*, *194*, 98–107. https://doi.org/10.1016/j.jep.2016.08.050

Mustapha, A. A. (2013). Ethno-medico-botanical uses of Securidaca longepedunculata Fresen (family-polygalaceae) from Keffi local government, Nasarawa state, Nigeria. *Journal of Natural Remedies*, *13*(2), 133–137.

Nadembega, P., Boussim, J. I., Nikiema, J. B., Poli, F., & Antognoni, F. (2011). Medicinal plants in Baskoure, Kourittenga Province, Burkina Faso: An ethnobotanical study. *Journal of Ethnopharmacology*, *133*(2), 378–395. https://doi.org/10.1016/j.jep.2010.10.010

Namukobe, J., Kiremire, B. T., Byamukama, R., Kasenene, J. M., Akala, H. M., Kamau, E., & Dumontet, V. (2015). Antiplasmodial compounds from the stem bark of *Neoboutonia macrocalyx* Pax. *Journal of Ethnopharmacology*, *162*(December), 317–322. https://doi.org/10.1016/j.jep.2014.12.018

Ngulde, S. I., Sandabe, U. K., Abounader, R., Dawson, T. K., Zhang, Y., Iliya, I., & Hussaini, I. M. (2019). Ethanol extract of *Securidaca longipedunculata* induces apoptosis in brain tumor (U87) cells. *BioMed Research International*, *2019*, 1–5. https://doi.org/10.1155/2019/9826590

Nguta, J. M., Mbaria, J. M., Gakuya, D. W., Gathumbi, P. K., & Kiama, S. G. (2010). Antimalarial herbal remedies of Msambweni, Kenya. *Journal of Ethnopharmacology*, *128*(2), 424–432. https://doi.org/10.1016/j.jep.2010.01.033

Nordor, V., Bellet, D., Siwo, G. H., & Siwo, G. H. (2018). Cancer – malaria: Hidden connections. *Open Biology*, *8*, 4–6.

Novotna, B., Polesny, Z., Pinto-Basto, M. F., Van Damme, P., Pudil, P., Mazancova, J., & Duarte, M. C. (2020). Medicinal plants used by "root doctors", local traditional healers in Bié province, Angola. *Journal of Ethnopharmacology*, *260*. https://doi.org/10.1016/J.JEP.2020.112662

Obasi, T. C., Benedec, D., Hanganu, D., Gheldiu, A. M., Vlase, L., Oniga, I., Pușcaș, C., Silaghi-Dumitrescu, R., & Oprean, R. (2020). Free radical scavenging activity and total polyphenol content of *Securidaca longipedunculata* roots and leaves extracts. *Farmacia*, *68*(1), 116–120. https://doi.org/10.31925/farmacia.2020.1.16

Ocan, M., Loyce, N., Ojiambo, K. O., Kinengyere, A. A., Apunyo, R., & Obuku, E. A. (2023). Efficacy of antimalarial herbal medicines used by communities in malaria affected regions globally: A protocol for systematic review and evidence and gap map. *BMJ Open*, *13*(7). https://doi.org/10.1136/bmjopen-2022-069771

Ochora, D. O., Kakudidi, E., Namukobe, J., Heydenreich, M., Coghi, P., Yang, L. J., Mwakio, E. W., Andagalu, B., Roth, A., Akala, H. M., Wong, V. K. W., & Yenesew, A. (2021). A new benzophenone, and the antiplasmodial activities of the constituents of *Securidaca longipedunculata* fresen (Polygalaceae). *Natural Product Research*, *0*(0), 1–9. https://doi.org/10.1080/14786419.2021.1925272

Ochora, D. O., Kakudidi, E. K., Namukobe, J., Ipulet, P., Wakoli, D. M., Okore, W., Mwakio, E. W., Yeda, R. A., Cheruiyot, A. C., Juma, D. W., Andagalu, B., Roth, A. L., Ogutu, B. R., Yenesew, A., & Akala, H. M. (2022). Synergism in antiplasmodial activities of artemether and lumefantrine in combination with *Securidaca longipedunculata*. *Plants*, *11*(47), 1–14. https://doi.org/10.3390/plants11010047

Ochora, D. O., Moodley, T., & Hayeshi, R. (2023). Phytochemistry, pharmacological activities, and drug interactions of pomegranate, *Punica granatum* L. (Punicaceae). In *Pomegranate – Biochemistry, Nutritional Benefits and Applications [Working Title]* (pp. 1–18). https://doi.org/10.5772/intechopen.1003779

Ochora, D. O., Saifudin Fidahusein, D., Nguta, J. M., & Akunda, E. M. (2014). Antimalarial activity and acute toxicity of four plants traditionally used in treatment of malaria in Msambweni District of Kenya. *European International Journal of Science and Technology*, *3*(7), 31–40. www.eijst.org.uk

Ohashi, M., Amoa-Bosompem, M., Kwofie, K. D., Agyapong, J., Adegle, R., Sakyiamah, M. M., Ayertey, F., Owusu, K. B. A., Tuffour, I., Atchoglo, P., Tung, N. H., Uto, T., Aboagye, F., Appiah, A. A., Appiah-Opong, R., Nyarko, A. K., Anyan, W. K., Ayi, I., Boakye, D. A., ... Ohta, N. (2018). In vitro antiprotozoan activity and mechanisms of action of selected Ghanaian medicinal plants against Trypanosoma, Leishmania, and Plasmodium parasites. *Phytotherapy Research*, *32*(8), 1617–1630. https://doi.org/10.1002/PTR.6093

Ojima, I., Lichtenthal, B., Lee, S., Wang, C., & Wang, X. (2016). Taxane anticancer agents: A patent perspective. *Expert Opinion on Therapeutic Patents*, *26*(1), 1–20. https://doi.org/10.1517/13543776.2016.1111872

Okoro, I. O., Osagie, A., & Asibor, E. O. (2010). Antioxidant and antimicrobial activities of polyphenols from ethnomedicinal plants of Nigeria. *African Journal of Biotecnology, 9*(20), 2989–2993. https://doi.org/10.5897/AJB09.1302

Opio, D., Andama, E., & Kureh, G. (2018). Ethnobotanical survey of antimalarial plants in areas of: Abukamola, Angeta, Oculokori and Omarari of Alebtong District in Northern Uganda. *European Journal of Medicinal Plants, 21*(4), 1–14. https://doi.org/10.9734/EJMP/2017/38043

Ouji, M., Augereau, J. M., Paloque, L., & Benoit-Vical, F. (2018). *Plasmodium falciparum* resistance to artemisinin-based combination therapies: A sword of Damocles in the path toward malaria elimination. *Parasite*, *25*. https://doi.org/10.1051/parasite/2018021

Ounjaijean, S., & Somsak, V. (2023). Synergistic antimalarial treatment of *Plasmodium berghei* infection in mice with dihydroartemisinin and *Gymnema inodorum* leaf extract. *BMC Complementary Medicine and Therapies*, *23*(1), 1–8. https://doi.org/10.1186/s12906-023-03850-y

Oyebode, O., Kandala, N. B., Chilton, P. J., & Lilford, R. J. (2016). Use of traditional medicine in middle-income countries: A WHO-SAGE study. *Health Policy and Planning*, *31*(8), 984–991. https://doi.org/10.1093/heapol/czw022

Quintana, M. del P., Smith-Togobo, C., Moormann, A., & Hviid, L. (2020). Endemic Burkitt lymphoma – an aggressive childhood cancer linked to *Plasmodium falciparum* exposure, but not to exposure to other malaria parasites. *APMIS*, *128*(2), 129–135. https://doi.org/10.1111/APM.13018

Raimi, I. O., Kopaopa, B. G., Mugivhisa, L. L., & Lewu, F. B. (2020). An appraisal of documented medicinal plants used for the treatment of cancer in Africa over a twenty-year period (1998–2018). *Journal of Herbal Medicine*, *23*, 1–18. https://doi.org/10.1016/j.hermed.2020.100371

Schultz, F., Osuji, O. F., Nguyen, A., Anywar, G., Scheel, J. R., Caljon, G., Pieters, L., & Garbe, L. A. (2021). Pharmacological assessment of the antiprotozoal activity, cytotoxicity and genotoxicity of medicinal plants used in the treatment of malaria in the greater Mpigi region in Uganda. *Frontiers in Pharmacology*, *12*(June), 1–20. https://doi.org/10.3389/fphar.2021.678535

Seca, A. M. L., & Pinto, D. C. G. A. (2018). Plant secondary metabolites as anticancer agents: Successes in clinical trials and therapeutic application. *International Journal of Molecular Sciences*, *19*(1). 263. https://doi.org/10.3390/ijms19010263

Siegel, R. L., Miller, K. D., & Jemal, A. (2019). Cancer statistics, 2019. *CA: A Cancer Journal for Clinicians*, *69*(1), 7–34. https://doi.org/10.3322/caac.21551

Song, X., Wei, W., Cheng, W., Zhu, H., Wang, W., Dong, H., & Li, J. (2022). Cerebral malaria induced by *Plasmodium falciparum*: Clinical features, pathogenesis, diagnosis, and treatment. *Frontiers in Cellular and Infection Microbiology, 12*(July), 1–14. https://doi.org/10.3389/fcimb.2022.939532

Tse, E. G., Korsik, M., & Todd, M. H. (2019). The past, present and future of anti-malarial medicines. *Malaria Journal, 18*(1), 1–21. https://doi.org/10.1186/s12936-019-2724-z

Veeresham, C. (2012). Natural products derived from plants as a source of drugs. *Journal of Advanced Pharmaceutical Technology and Research, 3*(4), 200–201. https://doi.org/10.4103/2231-4040.104709

Ward, M., & Benelli, G. (2016). Avian and simian malaria: Do they have a cancer connection? *Parasitology Research, 116*(3), 839–845. https://doi.org/10.1007/s00436-016-5352-3

WHO Global Malaria Programme. (2019). Artemisinin resistance and artemisinin-based combination therapy efficacy. https://iris.who.int/bitstream/handle/10665/274362/WHO-CDS-GMP-2018.18-eng.pdf (accessed 12 June 2025).

WHO International Agency for Cancer Research. (2024). www.iarc.who.int/ (accessed 2 June 2025).

WHO Guidelines for malaria. (2023). Instituitional respiratory for information sharing. https://iris.who.int/handle/10665/366432 (accesssed 12th June 2025).

World Health Organization (WHO). (2020). *World Health Organization Report on Cancer*. www.who.int/publications/i/item/who-report-on-cancer-setting-priorities-investing-wisely-and-providing-care-for-all (accessed 12th June 2025).

WHO World Malaria Report. (2024). www.who.int/teams/global-malaria-programme/reports/world-malaria-report-2024 (accessed 12th June 2025).

Williamson J. (1972). *Useful Plants of Malawi*. 1st Ed. Govt. Printer. ISBN: 1021079026950.

Wyss, K., Granath, F., Wångdahl, A., Djärv, T., Fored, M., Naucler, P., & Färnert, A. (2020). Malaria and risk of lymphoid neoplasms and other cancer: A nationwide population-based cohort study. *BMC Medicine, 18*(1), 1–13. https://doi.org/10.1186/s12916-020-01759-8

Yang, C., Wang, Z., Qiu, Y., Zha, H., & Yang, X. (2020). New hemiterpene and furolactone-type lignan glycosides from *Securidaca inappendiculata* Hassk. *Phytochemistry Letters, 37*(January), 42–46. https://doi.org/10.1016/j.phytol.2020.04.001

Yang, X. D., An, N., Xu, L. Z., & Shi-Lin, Y. (2002). New xanthone glycosides from *Securidaca inappendiculata*. *Journal of Asian Natural Products Research, 4*(2), 141–145. https://doi.org/10.1080/10286020290027434

Yuan, H., Ma, Q., Ye, L., & Piao, G. (2016). The traditional medicine and modern medicine from natural products. *Molecules, 21*(5), 559. https://doi.org/10.3390/molecules21050559

Zha, H., Wang, Z., Yang, X., Jin, D., Hu, L., Zheng, W., Xu, L., & Yang, S. (2015). New acylated triterpene saponins from the roots of *Securidaca inappendiculata* Hassk. *Phytochemistry Letters, 13*, 108–113. https://doi.org/10.1016/j.phytol.2015.05.022

Zhang, L.-J., Zhang, L., Xu, L.-Z., Yang, X.-D., & Zou, Z.-M.-L. Y. (2005). A new benzophenone from the root of *Securidaca inappendiculata*. *Acta Pharmaceutica Sinica*, *40*(6), 536–538.

Zuo, J., Mao, K. J., Yuan, F., Li, X., & Chen, J. W. (2014). Xanthones with anti-tumor activity isolated from *Securidaca inappendiculata*. *Medicinal Chemistry Research*, *23*(11), 4865–4871. https://doi.org/10.1007/s00044-014-1051-8

Chapter 2

Bioactive phytochemical compounds from the genus *Securidaca* L.

Salome Osunga,[1*] *Kennedy Olale,*[1] *Rael Masai,*[2] *and Douglas Ongeri Ochora*[2]
[1]Department of Chemistry, Kisii University, Kisii, Kenya
[2]Department of Biological Sciences, Kisii University, Kisii, Kenya

2.1 INTRODUCTION

Securidaca is a pantropical genus within the Polygalaceae family, comprising over 49 species native to tropical America, Asia, and Africa (Bojase and Ombito 2024). Several authors have documented varying numbers of species in this family, including 43 species (Zuo, Mao, et al. 2014), 49 (Bojase and Ombito 2024), 66 (Aymard-Corredor 2022), and about 80 (Yang, Xu, and Yang 2001; Mongalo et al. 2015). Plant species of this family have a long history of use in traditional medicine across diverse cultures for the treatment of a wide range of ailments, including inflammation and tuberculosis (Kang et al. 2005); pain (Ojewole 2008); snakebites (Das 2009); ascariasis, venereal diseases, and stomach ache (Bruschi et al. 2011); respiratory infections and malaria (Mongalo et al. 2015); toothache, infertility, headache, and cough (Kaniaru et al. 2021).

Plants in the genus *Securidaca* accumulate phytochemicals such as flavonoids (Hamburger, Gupta, and Hostettmann 1985), organic acids (Mahmood et al. 1993), sucrose derivatives (Tommasi et al. 1993), saponins (Timité et al. 2010), and alkaloids (Onobruche et al. 2017). These compounds have been reported to exhibit several biological activities, including antimicrobial, antioxidant, and anticancer activities (Bansal et al. 2019), and have been proposed as a solution to increasing antimicrobial resistance (Osunga et al. 2023). The genus *Securidaca* encompasses several species with diverse ethnomedicinal uses and pharmacological potential. Among the most mentioned species are *Securidaca longepedunculata, S. inappendiculata, S. welwitschiii, S. diversifolia, S. cacumina, S. marginata, S. savannarum, S. leiocarpa, S. planchoniana, S. calophylla, S. scandens, S. divaricata, S. lophosoma, S. micheliana, S. ovalifolia,* and *S. tomentosa,* although many are endemic to particular geographical areas (Aymard-Corredor 2022).

***Correspondence:** sosunga@kisiiuniversity.ac.ke/osungasally@gmail.com

 DOI: 10.1201/9781003544463-2

Several Polygalaceae species are significant producers of phytochemical compounds that are primarily used as analgesics and anti-inflammatory agents in traditional medicine. These include sterols, xanthones, flavonoids, phenols, alkaloids, derivatives of benzoic acid, terpenes, and glycosides (Yang et al., 2002; Tikisa et al. 2019). However, the biological activities of most of these phytochemicals are not well documented. Therefore, this study aimed to review and document the phytochemical constituents, pharmacology, ethnobotanical uses, and toxicity of bioactive compounds from the genus *Securidaca*.

2.2 MATERIALS AND METHODS

2.2.1 Systematic review method

A systematic review was conducted following the Preferred Reporting Items for Systematic Reviews and Meta-Analyses (PRISMA) guidelines (Page et al. 2021). Relevant information in English on the ethnopharmacological uses, phytochemicals, and bioactivities of the isolated compounds was obtained using various search engines, including Scopus, Google Scholar, Google, Springer, Elsevier, PubMed, Science Direct, ResearchGate, and Web of Science databases, with no restrictions on publication dates. The search terms included "*Securidaca* and its associated species", phytochemical compounds from *Securidaca*, biological activities of *Securidaca* species, and bioactive compounds of *Securidaca*. The chemical structures of the compounds were determined using ChemDraw Ultra 8.0. The PubChem and ChemSpider databases were used to verify the International Union of Pure and Applied Chemistry (IUPAC) names of isolated phytochemicals. The data included species name, solvents used for extraction, extracted phytochemical compounds, bioactivities of the phytochemical compounds, and the results of the bioactivities.

2.3 RESULTS AND DISCUSSIONS

2.3.1 The plants and biological activities

Plants of the genus *Securidaca* are known for their diverse metabolite profiles. The literature search revealed that the genus *Securidaca* has close to 80 species (Yang, Xu, and Yang 2001; Mongalo et al. 2015). The reported species are distributed across America, Asia, and Africa. An analysis of the literature revealed that *S. longepedunculata, S. inappendiculata, S. welwitschiii*, and *S. diversifolia* are the four major species which have been extensively investigated for phytochemical profiling, and their compounds have been isolated or purified. Other mentioned species with no phytochemical information include *S. cacumina, S. marginata, S. savannaru, S. leiocarpa, S. planchoniana, S. calophylla, S. scandens,*

S. divaricata, S. lophosoma, S. micheliana, S. paniculata, S. ovalifolia, and *S. tomentosa.*

S. longepedunculata and *S. inappendiculata* have the widest range of use and study. *S. longepedunculata* has been used traditionally in treating various ailments such as malaria, fever, headache, pain, stomach ache, constipation, hernias, ascariasis, coughs, pneumonia, tuberculosis, skin infections, rheumatism, epilepsy, abortion, sexually transmitted infections, as an aphrodisiac for men, and for tick prevention in animals (Mongalo et al. 2015). *S. inappendiculata* is commonly utilised for treating inflammations and rheumatoid arthritis (Yang, Xu, and Yang 2001). Extracts from *Securidaca* species have demonstrated various biological activities, as described in Sections 2.3.1.1 to 2.3.1.6.

2.3.1.1 Reducing and stabilising agents of silver nanoparticles (AgNPs)

To manage the characteristics of AgNPs and guarantee their stability for a variety of applications, stabilising and reducing agents are essential during the synthesis process. The reduction and stabilisation of silver nanoparticles using plant components provide an environmentally friendly, economical, and sustainable method for creating biocompatible nanomaterials with improved therapeutic efficacy.

In a study by Jayeoye et al. (2021), stem extracts of *S. inappendiculata* were used as reducing and stabilising agents of silver nanoparticles (SISE AgNPs) and were found to show antioxidative activity when compared to the ascorbic acid standard. This can be attributed to the presence of phenolic and other bioactive compounds in the plant extracts, which can donate hydrogen or protons to free radicals, thereby terminating the chain reaction of oxidation. Furthermore, SISE AgNPs exhibited potent antimicrobial capacity against some food-borne microbes by inhibiting both Gram-negative and Gram-positive bacteria. Additionally, SISE AgNPs conveniently serve as detection probes for the efficient monitoring of toxic heavy metals in aqueous environments. The fabricated SISE AgNPs were stable in an aqueous environment owing to the numerous SISE bioactive materials serving as stabilisers, and could detect Hg^{2+} in wastewater samples with good accuracy.

2.3.1.2 Anticancer and antidiabetics

For thousands of years, plants have been an essential part of traditional medicine because they provide a wealth of bioactive substances with therapeutic potential. Several plant species contain phytochemicals which exhibit potential as antidiabetic and anticancer agents, generally with fewer adverse effects than manufactured medications (Saha, Moitra, and Sanyal 2022; Hashim et al. 2025). This has prompted extensive scientific research and characterisation of bioactive compounds, offering the possibilities for the creation of innovative and supplementary treatment approaches.

An experiment using cervical cancer cells treated with extracts from the dried roots of *S. longepedunculata* showed significant cell alterations 48 h after treatment. This phenomenon may be related to autophagy and apoptosis. Two saponin fractions isolated from *S. longepedunculata* mediated apoptosis in cervical cancer cell lines (Obasi et al. 2018). The cytotoxic activity of the isolated saponins showed that the compounds had moderate activities against Lewis lung carcinoma cells, while another exhibited moderate cytotoxic activity against human breast cancer MCF-7 cells with an IC_{50} value of 47.93 μmol L (-1) (Zha et al. 2015).

The *in vitro* antidiabetic effects of AgNPs synthesised using 95% ethanol extract from the stem of *S. longepedunculata* (SISE) showed potent alpha-amylase and alpha-glucosidase inhibitory activity at concentrations ranging from 0.2 to 1.4 μg/mL. The percentage inhibition of alpha-amylase ranged between 5.3% and 100% at 0.2–1.4 μg/mL, while the percentage inhibition for alpha-glucosidase enzyme was between 9% and 100% (Jayeoye et al. 2021). To explore the glucose-lowering, insulinotropic, pancreatic-protective, and antioxidant effects of *S. inappendiculata* stem extract (SIE) in high fructose/STZ-induced diabetic rats, Olatunji et al. (2021) found that SIE administration attenuated diabetes-induced body weight loss, serum insulin decline, increased fasting blood glucose, lipid profiles, and liver and kidney dysfunction. SIE exhibited significant antidiabetic, antihyperlipidaemic, and antioxidant effects in diabetic rats by suppressing blood glucose levels, lipid peroxidation, lipid dysfunction, hepatorenal function, insulin secretion, and enzymatic antioxidant activity. Additionally, using dried *S. inappendiculata* stem samples, the neuroprotective effects of the *S. inappendiculata* polyphenol-rich extract (SiPE) in diabetic rats were evaluated. According to the findings, SiPE considerably reduced hyperglycaemia-induced memory and cognitive impairment in diabetic rats following eight weeks of treatment (Pang et al. 2021).

2.3.1.3 Anti-arthritis and anti-inflammatory activity

Traditional medical systems worldwide have long used plants to reduce inflammation and arthritis symptoms, and contemporary scientific research is still investigating and validating this practice. Numerous plants contain bioactive substances that have been shown to exert analgesic and anti-inflammatory effects, which may help treat certain ailments.

Experiments by Zhou et al. (2020), on the relationship between the therapeutic effect of Xanthose Rich Factor (XRF) from *S. inappendiculata* bark and collagen-induced arthritis in model rats, observed that XRF treatment markedly reduced the severity of the condition. According to this study, XRF's joint protective effects were brought about by blocking the glycolysis/

nicotinamide phosphoribosyltransferase (NAMPT) pathway, which in turn controlled macrophage polarisation. *S. inappendiculata* fatty oil (SIF) was found to have substantial antirheumatic activities (*in vivo)* and it alleviated local inflammation, reduced endogenous pathological factors, and restored biochemical parameters in adjuvant-induced arthritis (AA) male ICR mice. According to Zuo et al. (2014), Jiang et al. (2018), SIF can suppress the pathological functions of fibroblast-like synoviocytes and protect joints from destruction under AA conditions.

A further study by Zuo et al. (2014) found that the *S. innapendiculata* dichloromethane (SID) fraction, which exerted anti-inflammatory, analgesic, and immunodepressive activities *in vivo* and *in vitro* in male DBA/1 mice, was believed to be the most important active fraction from the plant for antirheumatic activity. Polyphenols are important anti-inflammatory constituents of *S. inappendiculata*. By acting as antioxidants, polyphenol-enriched fractions reduced oxidative stress and consequently inhibited NAMPT-mediated TLR4/NF-κB activation. This suggests that NAMPT is an important therapeutic target of polyphenols and that their antioxidant capacity is essential for controlling ROS-sensitive pro-inflammatory pathways (Ji et al. 2021).

2.3.1.4 Anti-epilepsy and anti-Alzheimer

Plant-derived chemicals may provide potential solutions to the intricate problems of epilepsy and Alzheimer's disease. Numerous plants have traditionally been used in traditional medical systems to treat neurological conditions, and current research is increasingly confirming these methods by identifying phytochemicals with neuroprotective and anticonvulsant properties.

Polyphenols isolated from dichloromethane and ethanol bark extracts of *S. longepedunculata* have been found to inhibit seizure-like paroxysms induced by acute pentylenetetrazole in larval zebrafish (Moussavi et al. 2024). These findings suggest that the extracts and their compounds could offer promising anticonvulsants and may have the potential to be utilised as a medicinal plant resource for the treatment of epilepsy (Moussavi et al. 2024).

Aqueous root extracts of *S. longepedunculata* have been shown to exhibit both anticholinesterase and antioxidant activities (Saliu and Olabiyi 2017). This plant showed potential in the management of Alzheimer's disease by exhibiting inhibitory activity on acetylcholinesterase and butyrylcholinesterase, which are linked to Alzheimer's disease. Therefore, the possible mechanism through which the extract exerts its neuroprotective properties may be the inhibition of cholinesterase activity and the prevention of oxidative stress-induced neurodegeneration (Saliu and Olabiyi 2017).

2.3.1.5 Anti-trypanosomal and antiplasmodial activity

Devastating illnesses such as malaria and trypanosomiasis, caused by the protozoan parasites *Plasmodium* and *Trypanosoma*, respectively, continue to be major global health concerns, especially in tropical and subtropical areas. The rise of drug-resistant parasite strains and the toxicity of certain synthetic substances highlight the urgent need for innovative treatment strategies, even though conventional pharmacological therapies are available. Many plant species have shown encouraging anti-trypanosomal and antiplasmodial properties, providing possible pathways for the creation of novel, safer, and more effective medicines (Kamanzi et al. 2004).

In vivo tests of the root extract of *S. longepedunculata* reduced the average parasitaemia of six *Mastomys coucha* mice by 48% (Aderbauer et al. 2008). The methanol root extract-high trypanocidal activity observed *for S. longepedunculata* caused complete cessation of motility of both *Trypanosoma brucei brucei* and *Trypanosoma congolense* within 60 min (Atawodi et al. 2003). The roots, leaves, and stems of *S. longepedunculata* extracts combined with artemether and lumefantrine showed synergistic *in vitro* and *ex vivo* antiplasmodial activities (Ochora et al. 2022). The root extract was the most potent and was enhanced when combined with artemether. Extracts from *S. longepedunculata* showed antiplasmodial activity against reference clones and clinical isolates using the SYBR Green I method (Ochora et al. 2021).

2.3.1.6 Crop pest control

Concerns regarding the sustainability and environmental effects of synthetic pesticides have led to a growing interest in the long-standing practice of using plants and their extracts to control crop pests in modern agriculture. To protect themselves from herbivorous insects and other pests, many plant species naturally produce a wide range of bioactive chemicals with insecticidal, repellant, or antifeedant properties. By treating crop pests in a potentially safer and more sustainable manner, these botanical pesticides support agricultural productivity and ecological balance (Kerebba et al. 2019).

The powdered dry root bark of *S. longepedunculata* was mixed with maize and cowpea and effectively reduced the numbers of *Sitophilus zeamais* and *Callosobruchus maculatus* emerging from these commodities, respectively, for more than nine months after treatment. This effect was reciprocated in grains treated with a methanol extract of the root bark, indicating that there were compounds that were either oviposition deterrents or directly toxic to adults or larvae (Stevenson et al. 2009). This effect was attributed to the presence of two novel bisdesmosidic saponins.

2.3.2 Bioactive phytochemicals

Several compounds have been identified in *Securidaca*, including sterols, xanthones, flavonoids, phenols, alkaloids, benzoic acid derivatives, terpenes, and glycosides (Yang et al. 2002; Tikisa et al. 2019). The secondary metabolites are shown in Figures 2.1–2.5. The most mentioned species in the literature for bioactive phytochemicals include *S. longepedunculata, S. inappendiculata, S. welwitschii,* and *S. diversifolia.*

2.3.2.1 Securidaca longepedunculata

Several benzoic acid derivatives (Figure 2.1) have been reported from chloroform root bark and methanol roots, including methyl-2-hydroxybenzoate (methyl salicylate) (**1**) and methyl-2-hydroxy-6-methoxy benzoate (**2**) (Jayasekara et al. 2002; Lognay et al. 2000). It has been suggested that 2 Methyl-2-hydroxybenzoate exhibits insecticidal activity against stored-grain insects. Benzyl 2-hydroxy-5-methoxy benzoate (**3**) has also been

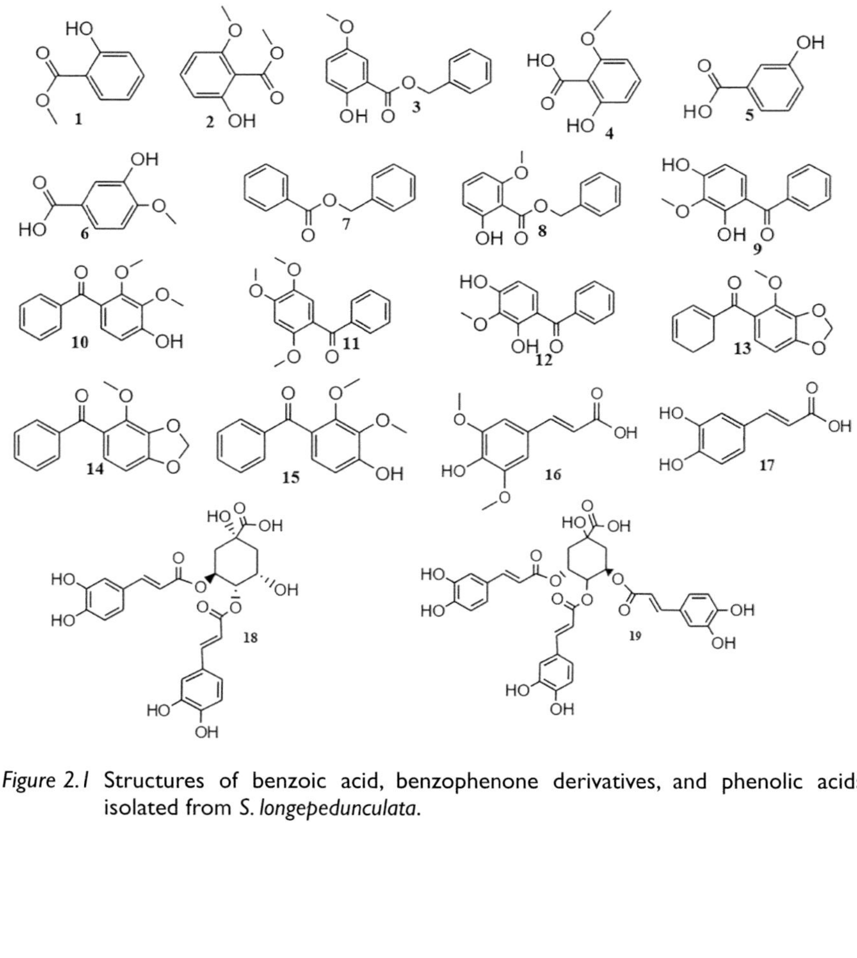

Figure 2.1 Structures of benzoic acid, benzophenone derivatives, and phenolic acids isolated from *S. longepedunculata*.

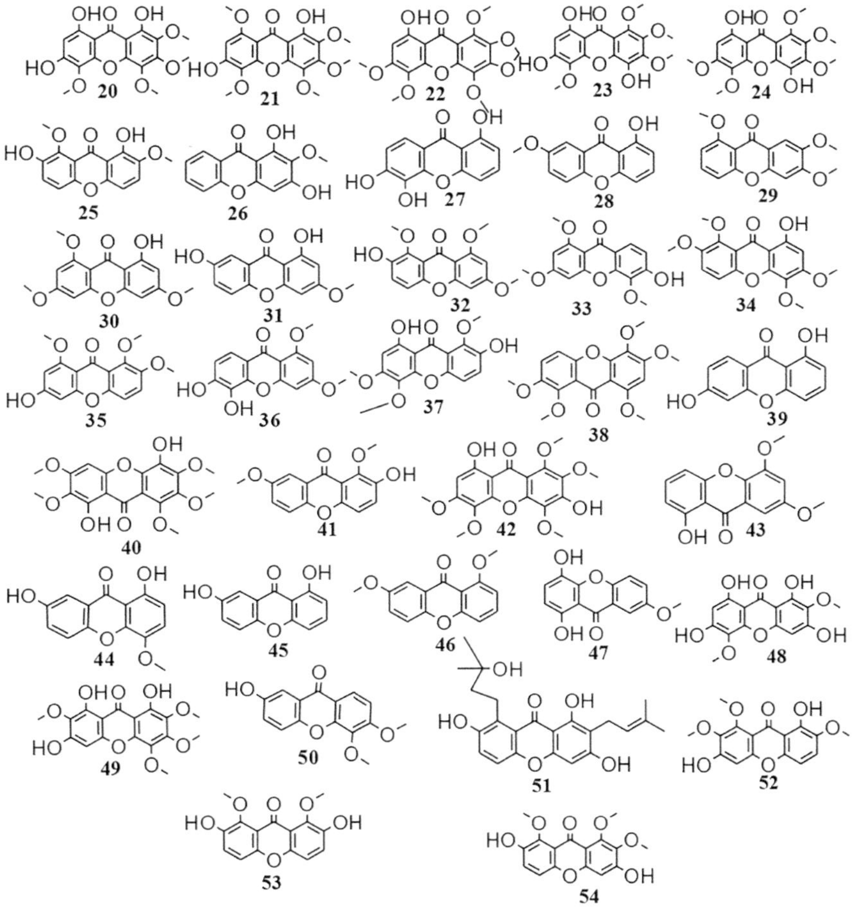

Figure 2.2 Structures of xanthones from *S. longepedunculata.*

isolated from methanolic root and acetone root bark extracts (Jayasekara et al. 2002; Keshebo and Choudhury, 2015).

Other benzoic acid derivatives, such as 6-hydroxy-2-methoxy-benzoic acid (**4**), have been isolated from chloroform and dried ethyl acetate roots (Lannang et al. 2006; Dibwe et al. 2013), while 3-hydroxy benzoic acid (**5**) and 3-hydroxy-4-methoxy benzoic acid (**6**) were isolated from the acetone root bark extract (Tikisa and Abdissa, 2019). Benzyl benzoate (**7**) and benzyl-2-hydroxy-6-methoxy-benzoate (**8**) were also isolated from the dried bark extract, and compound **8** was reported to be one of the most toxic compounds assessed in the assay for maximum tolerated concentration (MTC) (Moussavi et al. 2024).

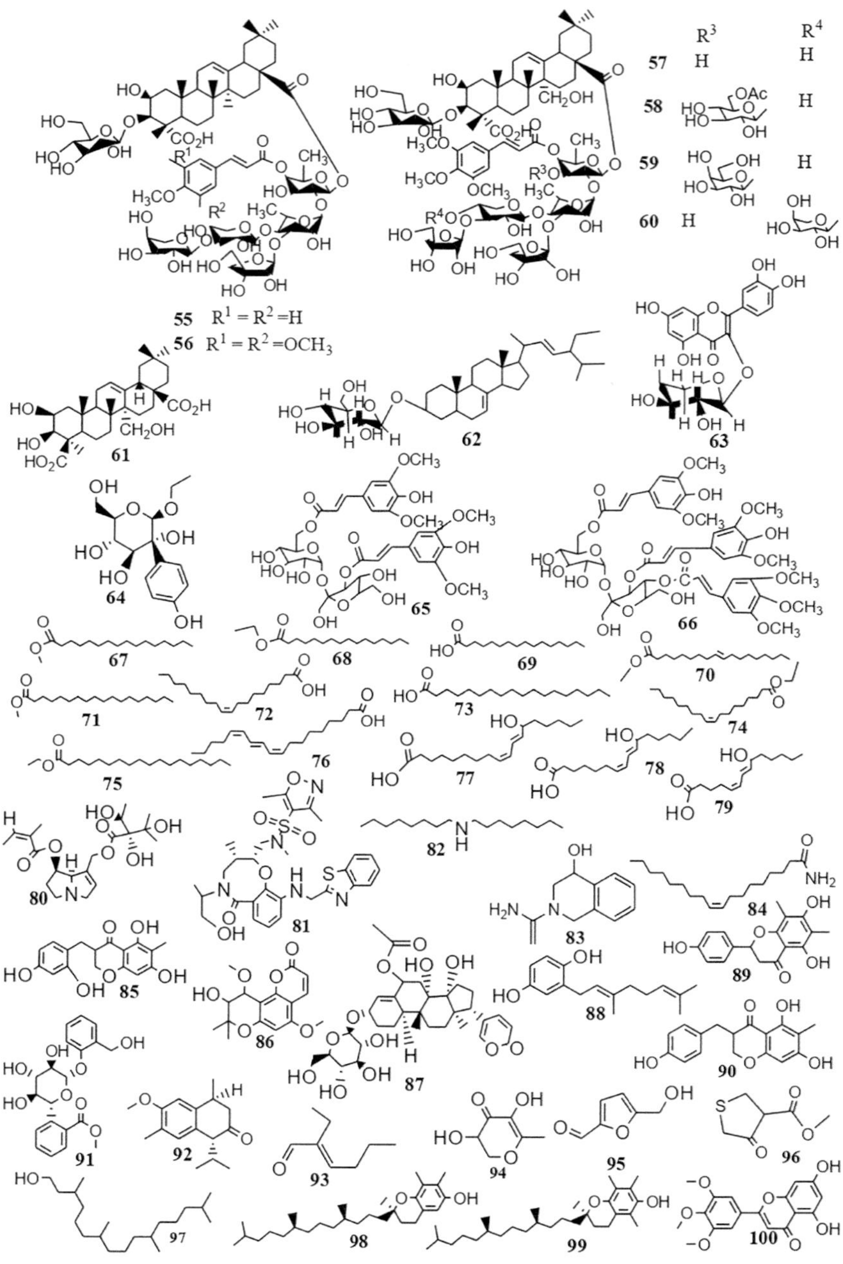

Figure 2.3 Structures of saponins, glycosides, sucrose derivatives, fatty acids, alkaloids, and other compounds from *S. longepedunculata.*

Compound	R^1	R^2	R^3	R^4	R^5	R^6	R^7	R^8
101	OH	OCH_3	H	H	H	H	H	OH
102	OCH_3	OCH_3	H	H	H	H	OH	H
103	OCH_3	OCH_3	H	H	H	H	OH	H
104	OH	H	H	H	H	H	H	OH
105	OCH_3	OCH_3	OH	H	OCH_3	H	H	OCH_3
106	OH	OCH_3	OH	H	H	H	H	OH
107	OH	H	OH	OCH_3	H	H	H	OH
108	OH	OCH_3	OH	H	H	OH	OCH_3	H
109	OH	OH	H	H	H	OCH_3	OCH_3	OCH_3
110	OH	OCH_3	OH	H	H	H	OH	OCH_3
111	H	H	OCH_3	OH	H	H	OCH_3	H
112	OH	H	OH	H	H	H	OH	H
113	OCH_3	OCH_3	H	H	H	H	OH	H
114	OH	OCH_3	OH	H	H	H	OH	H
115	OH	H	OCH_3	OCH_3	H	H	OH	H
116	OH	OCH_3	OH	H	H	OH	OCH_3	H
117	OCH_3	H	OH	OCH_3	H	H	H	OH

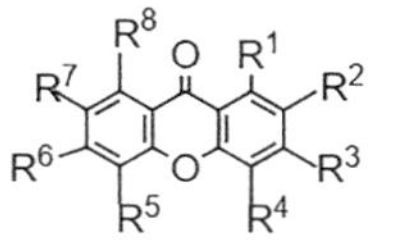

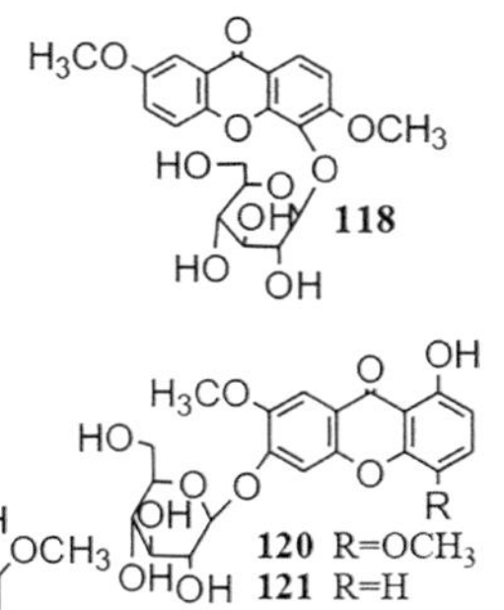

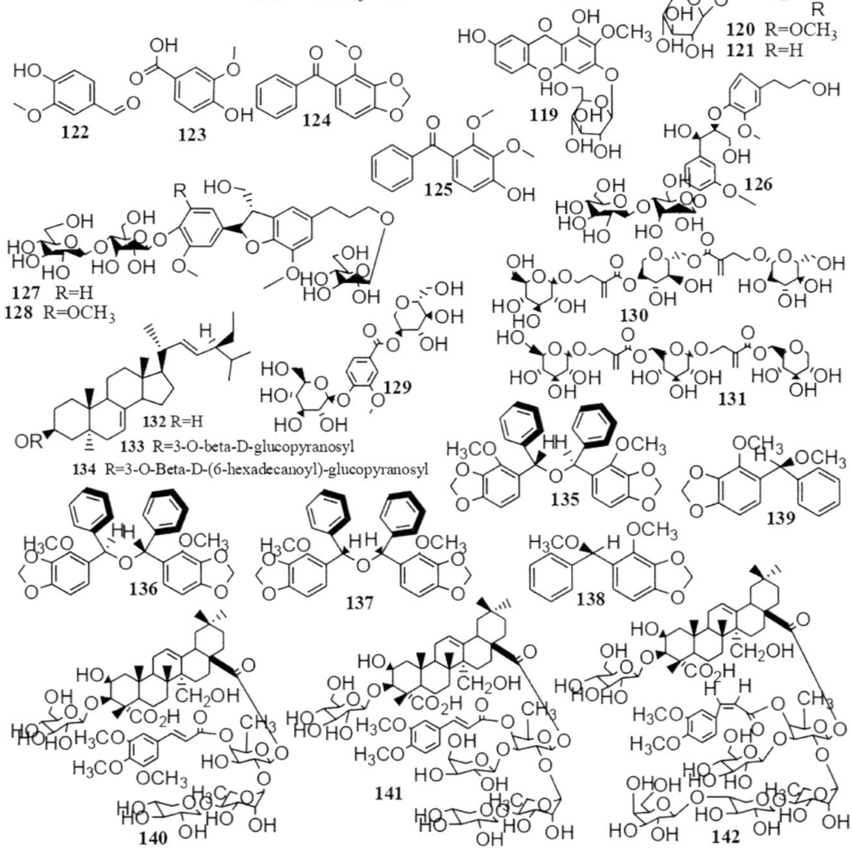

Figure 2.4 Structures of compounds isolated from *S. inappendiculata*.

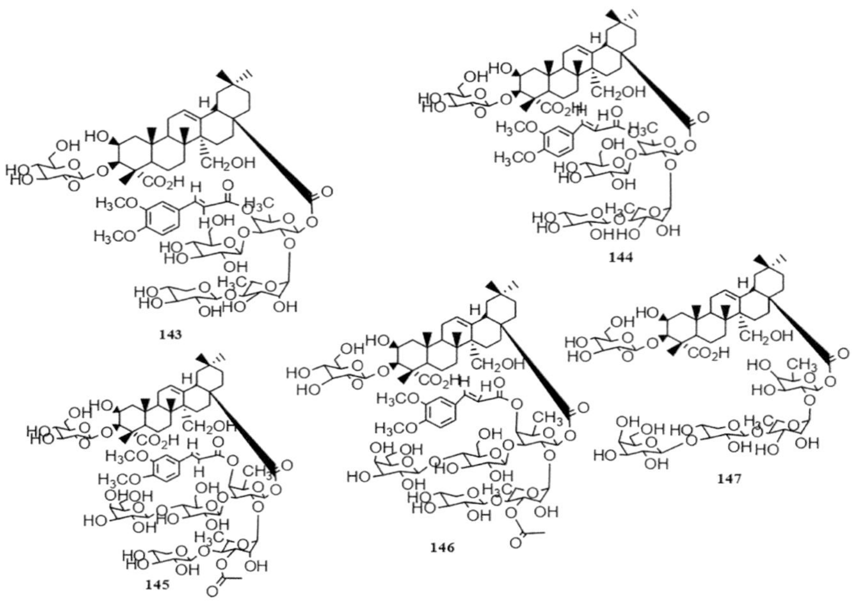

Figure 2.5 Structures of compounds isolated from *S. welwitschii.*

Benzophenone derivatives (Figure 2.1), 2,4-dihydroxy-3'-methoxy benzophenone (**9**), 2,3-dimethoxy-4 hydroxy benzophenone (**10**), 2,4,5-tri methoxy- benzophenone (**11**), 2,4-dihydroxy-3-methoxy benzophenone (**12**), and 2-methoxy-3,4-methylenedioxy benzophenone (**13**), were isolated from chloroform root extract (Dibwe et al. 2013).

Other benzophenone derivatives from this plant include (4-methoxy-benzo [1,3] dioxol-5-yl)-phenylmethanone (**14**) from dichloromethane-dried roots, which exhibited antibacterial activity against *Pseudomonas aeruginosa* and antifungal activity against *Aspergillus fumigatus* (Joseph et al. 2006). Moreover, 2,3-dimethoxy-4-hydroxy- benzophenone (**15**) was isolated from dichloromethane dried bark extract (Moussavi et al. 2024).

The methanolic extract of *S. longepedunculata* bark resulted in the isolation of phenolic acids (Figure 2.1): sinapoic acid (**16**), caffeic acid (**17**), 3,4,5-tricaffeoyl-D-quinic acid (**18**), and 4,5-dicaffeoyl-D-quinic acid (**19**) (Tommasi et al., 1993).

Numerous xanthones have been isolated from *S. longepedunculata* (Figure 2.2). The chloroform root extract of *S. longepedunculata* (Dibwe et al. 2013) yielded: 1,6,8-trihydroxy-2,3,4,5-tetramethoxyxanthone (**20**), 1,6-dihydroxy-2,3,4,5,8-pentamethoxyxanthone (**21**), 8-hydroxy-1,4,5,6-tetramethoxy-2,3-methylenedioxyxanthone (**22**), 4,6,8-trihydroxy- 1,2,3,5-tetramethoxyxanthone (**23**), 4,8-dihydroxy-1,2,3,5,6-

pentamethoxyxanthone (**24**), 1,7-dihydroxy-2,8- dimethoxyxanthone (**25**), 1,3-dihydroxy-2-methoxyxanthone (**26**), 1,5,6-trihydroxyxanthone (**27**), 1-hydroxy-7-methoxyxanthone (**28**), 2,3,8-trimethoxyxanthone (**29**), 1-hydroxy-3,6,8-trimethoxyxanthone (**30**), 1,7-dihydroxy-3-methoxyxanthone (**31**), 7-hydroxy-1,3,8-trimethoxyxanthone (**32**), 3-hydroxy-4,6,8-trimethoxyxanthone (**33**), 1-hydroxy-3,4,7,8-tetramethoxyxanthone (**34**), 6-hydroxy-1,2,8-trimethoxyxanthone (**35**), 5,6-dihydroxy-1,3-dimethoxyxanthone (**36**), 2,8-dihydroxy-1,5,6-trimethoxyxanthone (**37**), 1,2,5,6,8-pentamethoxyxanthone (**38**), and 1,6-dihydroxyxanthone (**39**). Compounds **23** and **24** demonstrated strong preferential cytotoxicity (PC_{50}) of 22.8 and 17.4 l M, respectively, by causing cell death in PANC-1, human pancreatic cancer cells in nutrient-deprived medium (Dibwe et al. 2013).

Dried ethyl acetate roots led to the isolation of xanthones: 1,5-Dihydroxy-2,3,6,7,8 pentamethoxyxanthone (**40**) and 2-Hydroxy-1, 7-dimethoxyxanthone (**41**) (Lannang et al. 2006). The ethyl acetate extract of the root bark yielded 3,8-dihydroxy- 1,2,4,5,6-pentamethoxyxanthone (**42**), 1-hydroxy-5,7-dimethoxyxanthone (**43**), 1,7-dihydroxy-4- methoxyxanthone (**44**), 1,7-dihydroxyxanthone (**45**) 1,7- dimethoxyxanthone (**46**), and 1,4-dihydroxy-7-methoxyxanthone (**47**) (Wedajo et al. 2022). Compounds **44** and **46** exhibited significant cytotoxic activity against the human cervical cancer cell line KB-3-1, with compound **44** demonstrating a higher IC_{50} value than griseofulvin, a reference drug. Compound **40** also exhibited antibacterial activity against *S. aureus* and antifungal activity against *Aspergillus fumigatus* and *Aspergillus niger* (Joseph et al. 2006).

Another study examined the *in vitro* activity on rabbit corpus cavernosum of 1,7-dihydroxy-4-methoxyxanthone (**44**) and 1,7-dimethoxy-2-hydroxyxanthone (**41**), isolated from roots. Compound **44** relaxed the corpus cavernosum by 63% at 1.8×10^{-5} mg/mL, while compound **41** showed no activity (Rakuambo, Meyer, and Hussein 2004). Xanthones 1,3,6,8-tetrahydroxy-2,5-dimethoxyxanthone (**48**) and 1,6,8-trihydroxy-2,3,4,7-tetramethoxyxanthone (**49**) were also obtained from the fresh root bark extract in acetone (Meyer et al. 2008). It was further reported that at 1.8×10^{-5} mg/mL, compounds **48** and **49** relaxed the smooth muscle of the corpus cavenosum by 97% and 30.5%, respectively, in contrast to the positive control (Viagra), which relaxed the muscles by 100.0%.

3, 4-Dimethoxy-7- hydroxyxanthone (**50**), isolated from the acetone root bark extract, exhibited weak antibacterial activity against *Staphylococcus aureus*, *Escherichia coli*, and *Pseudomonas aeruginosa,* and significant activity against *Bacillus subtilis*, which was equivalent to gentamicin, the reference drug (Tikisa, Abdissa, and Abdissa 2019).

1,3,7-Trihydroxy-2-(3-methyl-2-butenyl)-8-(3-hydroxy-3-methylbutyl) xanthone (**51**) was obtained from an ethyl acetate-dried stem bark extract (Abubakar et al. 2022a; Abubakar et al. 2022b). In addition, some xanthones were identified in the dichloromethane-dried bark extract:

1,6-dihydroxy- 2,7,8-trimethoxyxanthone (**52**), 2,7-dihydroxy-1,8-dimethoxyxanthone (**53**), and 3,7-dihydroxy-1,2,8-trimethoxyxanthone (**54**). According to the MTC assay, compounds **35** and **42** were the most toxic, while compound **44** was one of the compounds that most significantly reduced paroxysms by changing the locomotor behaviour of the γ-amino butyric acid A ($GABA_A$) receptor antagonist pentylenetetrazole (PTZ), which caused seizures in larval zebrafish (Moussavi et al. 2024).

Several saponins (Figure 2.3) have been isolated from the roots of *S. longepedunculata*. The methanolic root bark yielded two bidesmodic saponins, securidacaside A (**55**) and securidacaside B (**56**) (Stevenson et al. 2009). Securidacaside A was found to be both toxic and deterrent to *Sitophilus zeamais* and *Callosobruchus maculates*. Additionally, four triterpene saponins have been obtained from 70% methanolic root extract: 3-O-β-D glucopyranosylpresenegenin 28-O-β-D-apiofuranosyl-(1→3)-β–D-pyranosyl-(1→4)-[β-D-apiofuranosyl-(1→3)]-α-L-rhamnopyranosyl-(1→2)-{4-O-[(E)-3,4,5-trimethoxycinnamoyl]}-β-D-fucopyranosyl ester (**57**), 3-O-β-D-glucopyranosylpresenegenin 28-O-β-D-apiofuranosyl-(1→3)-β-D-xylopyranosyl-(1→4)-[β-D-apiofuranosyl-(1→3)]-α-L-rhamnopyranosyl-(1→2)-[(6-O-acetyl)-β-D-glucopyranosyl-(1→3)]-{4-O-[(E)-3,4,5-trimethoxycinnamoyl]}-β-D-fucopyranosyl ester (**58**), 3-O-β-D-glucopyranosylpresenegenin 28-O-β-D-apiofuranosyl-(1→3)-β-D-xylopyranosyl-(1→4)-[-β–D-apiofuranosyl-(1→3)]-α-L-rhamnopyranosyl-(1→2)-[β-D-galactopyranosyl-(1→3)]-{4-O-[(E)-3,4,5-trimethoxycinnamoyl]}-β-D-fucopyranosyl ester (**59**), and 3-O- β-D-glucopyranosylpresenegenin 28-O-β-D-apiofuranosyl-(1→3)-[α-L-arabinopyranosyl-(1→4)]-β-D-xylopyranosyl-(1→4)-[β-D-apiofuranosyl-(1→3)]-α-L-rhamnopyranosyl-(1→2)-{4-O-[(E)-3,4,5-trimethoxycinnamoyl]}-β-D-fucopyranosyl ester (**60**) (Mitaine-Offer et al. 2010). It has been shown that aqueous roots contain presenegenin (**61**) (Mongalo et al. 2015; Abubakar et al., 2019).

The bark and stem bark were used to extract glycosides (Figure 2.3): Δ^7Stigmasterol-3-O-β-D-glucopyranoside (**62**) and quercetin-3-O-β-D-xyloside (**63**), respectively (Debella et al. 2000), and 2-(4-Hydroxyphenyl) ethyl beta-D-glucopyranoside (**64**) was extracted from the ethyl acetate-dried stem bark extract (Abubakar et al. 2022a; Abubakar et al. 2022b).

The sucrose derivatives β-D-(3,4-disinapoyl)fructofuranosyl-α-D(6-sinapoy1)glucopyranoside (**65**) and β-D-(3-sinapoyl) fructofuranosyl-α-D (6-sinapoyl)glucopyranoside (**66**) were obtained from methanolic bark (Tommasi, Piacente, De Suisone, 1993).

Ethanolic dried leaves *of S. longepedunculata* yielded fatty acids (Figure 2.3): methyl palmitate (**67**), ethyl palmitate (**68**), palmitic acid (**69**), methyl (E)-octadec-9-enoate (**70**), methyl stearate (**71**), oleic acid (**72**), stearic acid (**73**), ethyl oleate (**74**), and ethyl stearate (**75**), whereas

9Z,11E,13Z-octadeca-9,11,13-trienoic acid (**76**) was isolated from ethyl acetate-dried stem bark extract (Egga et al. 2023; Abubakar et al. 2022a; Abubakar et al. 2022b). Additionally various fatty acids have been yielded from light petroleum (b.p 31–51°C) seed extract: (9Z, 11E)-13-hydroxyoctadeca-9,11-dienoic acid (**77**), (7Z, 9E)-11-hydroxyhexadeca-7,9-dienoic acid (**78**), and (5Z, 7E)-9-hydroxytetradeca-5,7-dienoic acid (**79**) (Smith Jr, Madrigal, and Plattner 1979).

Various alkaloids (Figure 2.3), [(7R,8R)-7-[(Z)-2-methylbut-2-enoyl]oxy-5, 6,7,8-tetrahydro-3H-pyrrolizin-1-yl]methyl (2R)-2,3-dihydroxy-2-[(1S)-1-hydroxyethyl]-3-methylbutanoate (**80**), N-[(2R,3S)-2-[[(3,5-dimethyl-4-isoxazolyl)sulfonyl-methylamino]methyl]-5-[(2R)-1-hydroxypropan-2-yl] -3-methyl-6-oxo-3,4-dihydro-2H-1,5-benzoxazocin-10-yl]-1,3-benzothiazole-2-carboxamide (**81**), N-octyloctan-1-amine (**82**), 4-hydroxy-3,4-dihydro-1H-isoquinoline-2-carboximidamide (**83**), and (Z)-Octadec-9-enamide (**84**), were isolated from the ethyl acetate-dried stem bark extract (Abubakar et al. 2022a; Abubakar et al. 2022b).

Other compounds (Figure 2.3) yielded from ethyl acetate-dried stem bark extract of *S. longepedunculata* were as follows: 5,7-dihydroxy-6-methyl-3-(2',4'-dihydroxybenzyl)-chroman-4-one(**85**),trans-O-methylgrandmarin(**86**), scilliroside (**87**), 2-[(2E)-3,7-dimethylocta-2,6-dien-1-yl] benzene-1,4-diol. (**88**), 5,7-dihydroxy-2-(4-hydroxyphenyl)-6,8-dimethyl-3,4-dihydro-2H-1-benzopyran-4-one (flavanoid) (**89**), 5,7-dihydroxy-6-methyl-3-(4-hydroxy benzyl)-chroman-4-one (**90**), [(2R,3S,4S,5R,6S)-3,4,5-trihydroxy-6-[2-(hydroxymethyl)phenoxy]oxan-2-yl]methylbenzoate (**91**), and (1S,4R)-7-methoxycalamenen-3-one (**92**) (Abubakar et al. 2022a; Abubakar et al. 2022b).

Additionally, ethanolic dried leaves yielded (Figure 2.3) 2-ethyl-2-hexenal (**93**), 2,3-dihydro-3,5-dihydroxy-6-methyl-4H-Pyran-4-one (**94**), 5-Hydroxymethylfurfural (**95**), 3-thiophenecarboxylic acid tetrahydro-4-oxo-methyl (**96**), phytol (terpene) (**97**), ɤ-tocopherol (**98**), 5,7-Dihydroxy-2-(3,4,5-trimethoxyphenyl)-4H-1- benzopyran-4-one (**99**), and ɤ-tocopherol (**100**) (Egga et al. 2023).

2.3.2.2 *Securidaca inappendiculata*

Previous studies on dried root extracts (95% ethanol) have indicated the presence of xanthones, terpenoids, polyphenols, steroids, lignans, and fatty acids (Zuo et al. 2014; Kang and Xu 2008). The identified xanthones are (Figure 2.2): 1,7-dihydroxy xanthone (**45**) and 1,7-dimethoxy xanthone (**46**), (Figure 2.3): 3,7-dihydroxy-1,2,8-trimethoxyxanthone (**55**) (Figure 2.4): 1,7-dihydroxy-2-methoxy xanthone (**101**); 7-hydroxy-1,2-dimethoxyxanthone (**102**), 3-hydroxy-4,6-dimethoxyxanthone (**103**), and 1,8-dihydroxyxanthone (**104**) (Zhang et al. 2005; Kang and Xu 2008).

Additionally, the dried stem ethanol extract yielded xanthones (Figure 2.2): 2-hydroxy-1,7-dimethoxyxanthone (**41**), 1,7-dihydroxy-4-methoxyxanthone (**44**), 1,7-dihydroxyxanthone (**45**); (Figure 2.4): 3-hydroxy-1,2,5,8-tetramethoxyxanthone (**105**), 1,3,8-trihydroxy-2-methoxyxanthone (**106**), 1,3,8-trihydroxy-4-methoxyxanthone (**107**), 1,3,6-trihydroxy-2,7-dimethoxyxanthone (**108**), 1,2-dihydroxy-6,7,8-trimethoxyxanthone (**109**), 1,3,7-trihydroxy-2,8-dimethoxyxanthone (**110**), 3,7-dimethoxy-4-hydroxyxanthone (**111**), 1,3,7-trihydroxyxanthone (**112**), 7-hydroxy-1,2-dimethoxyxanthone (**113**), 1,3,7-trihydroxy-2-methoxyxanthone (**114**), 1,7-dihydroxy-3,4-dimethoxyxanthone (**115**), 1,3,6-trihydroxy-2,7-dimethoxyxanthone (**116**), and 3,8-dihydroxy-1,4-dimethoxyxanthone (**117**) (Yang, Xu, and Yang 2001; Zuo, et al. 2014). Xanthones, 3-hydroxy-1,2,5,8-tetramethoxyxanthone (**105**), 1,3,8-trihydroxy-2-methoxyxanthone (**106**), 1,3,8-trihydroxy-4-methoxyxanthone (**107**), and 1,3,6-trihydroxy-2,7-dimethoxyxanthone (**116**), have been isolated from both the stems and roots (Kang, Ji, and Wang 2009).

Xanthone glycosides and glycosides (Figure 2.4), securixanside A (**118**), securixanside B (**119**) 6-O-β-D-glucopyranosyl- 1-hydroxy-4,7-dimethoxyxanthone (securixanside C) (**120**), and 6-O-β-D-glucopyranosyl-1-hydroxy-7-methoxyxanthone (**121**), have been isolated from 95% ethanolic extract of the dried stem (Yang et al. 2002).

Benzoic acid derivatives (Figure 2.4), 4-hydroxy-3-methoxybenzaldehyde (**122**) and 4-hydroxy-3-methoxybenzoic acid (**123**) were isolated from chloroform root extract (Kang et al. 2005). In addition, benzophenone derivatives (Figure 2.4), 2-methoxy-3,4-(methylenedioxy)benzophenone (**124**) and 4-hydroxy-2,3-dimethoxybenzophenone (**125**) have also been isolated from the chloroform root extract (Kang et al. 2005).

Glycosides (Figure 2.4), (7R,8S)-4-hydroxy-3,3′-dimethoxy-8,4′-oxyneoligna-7,9,9′-triol-4-O-β-D-glucopyranosyl-(1→4)-β-D-glucopyranoside (**126**), (7R,8S)-4-hydroxy-3,5′-dimethoxy-4′,7-epoxy-8,3′-neoligna-9,9′-diol-9’-O-β-D-glucopyranosyl-4-O-[β-D-glucopyranosyl-(1→4)]-β-D-glucopyranoside (**127**), (7R,8S)-4-hydroxy-3,5,5′-trimethoxy-4′,7-epoxy-8,3′-neoligna-9,9′-diol-9’-O-β-D-glucopyranosyl-4-O-[β-D-glucopyranosyl-(1→4)]-β-D-glucopyranoside (**128**), securiphenoside B (**129**), [6-O-(2-methylenebutanoyl-4-O-β-D-glucopyranosyl)-1,5-anhydroglucitol-(2→1)]-2-methylenebutanoate-4-O-β-D-glucopyranoside (securiterpenoside E) (**130**), and [1-O-(2-methylenebutanoyl-1-O-1,5-anhydroglucitol)-β-D-glucopyranosyl-(6→1)]-2-methylenebutanoate-4-O-βD-glucopyranoside (securiterpenoside F) (**131**), have been isolated from ethanolic dried stems of *S. inappendiculata*. Compound **132** exhibited moderate hepatoprotective activity (Ji et al. 2019).

Sterol glycosides and sterols (Figure 2.4), including spinasterol (**132**), 3-O-b-D-glucopyranosyl-spinasterol (**133**), and securisterol (**134**), were isolated from 95% ethanolic dried roots (Zhang et al. 2005). Diarymethane derivatives (Figure 2.4), including securine A (**135**), securine B (**136**), securine C (**137**), securine D (**138**), and securine E (**139**), were obtained from a 95% ethanolic stem extract. Compounds (**135**), (**136**), and (**137**) exhibited modest antiplasmodial activity, with corresponding IC_{50} values of 0.9, 1.4, and 1.5 µM, respectively (Zhou et al. 2020).

Saponins (Figure 2.4), securioside C (**140**), securioside D (**141**), and (Z)–isomer securioside E (**142**), were extracted from the 95% ethanol (EtOH) roots with compounds (**140**) and (**141**) showing slight cytotoxic activity against Lewis lung carcinoma (LLC) cells with an IC_{50} values of 45.56 and 85.98 mM, respectively (Zha et al. 2015).

2.3.2.3 *Securidaca welwitschii*

Phytochemical investigations of *S. welwitschii* stem bark methanol (MeOH) extract showed the presence of triterpene saponins and a sucrose ester (Timité et al. 2010). These compounds (Figure 2.5) included 3-O-(β-D-glucopyranosyl)presenegenin 28-{O-β-D-xylopyranosyl-(1→4)-O-α-L-rhamnopyranosyl-(1→2)-O-[β-D-glucopyranosyl-(1→3)]-4-O-[(E)-3,4-dimethoxycinnamoyl]-β-D-fucopyranosyl}ester (**143**), 3-O-(β-D-glucopyranosyl)presenegenin-28-{O-β-D-xylopyranosyl-(1→4)-O-α-L-rhamnopyranosyl-(1→2)-O-[β-D-glucopyranosyl-(1→3)]-4-O-[(E)-3,4-dimethoxycinnamoyl]-β-D-fucopyranosyl}ester(**144**), 3-O-(β-D-glucopyranosyl)presenegenin-28-{O-β-D-galactopyranosyl-(1→4)-O-β-D-xylopyranosyl-(1→4)-O-3-O-acetyl-α-L-rhamnopyranosyl-(1→2)-O-[β-D-glucopyranosyl-(1→3)]-4-O-[(E)-3,4-dimethoxy-cinnamoyl]-β-D-fucopyranosyl}ester (**145**), 3-O-(β-D-glucopyranosyl) presenegenin-28-{O-β-D-galactopyranosyl-(1→4)-O-β-D-xylopyranosyl-(1→4)-O-3-O-acetyl-α-L- rhamnopyranosyl-(1→2)-O-[β-D-glucopyranosyl-(1→3)]-4-O-[(Z)-3,4-dimethoxy- cinnamoyl]-β-D-fucopyranosyl}ester (**146**), and (3-O-[(Z)-3,4,5-trimethoxycinnamoyl]-β-D-fructofuranosyl 6-O-[(E)-3,4,5-trimethoxycinnamoyl]-α-D-glucopyranoside (**147**) (Timité et al. 2010).

2.3.2.4 *Securidaca diversifolia*

Methanoilic dried leaves extract of *S. diversifolia* led to isolation of (Figure 2.6): quercetin-3-sophoroside (**148**), quercetin-3-galactoside (hyperoside) (**149**), isoquericitrin (**150**), guaijaverin (**151**), reynoutrin (**152**), kaempferol-3-α-arabinopyranoside (**153**), and quercitrin (**154**) (Hamburger, Gupta, and Hostettmann 1985).

2.4 CONCLUSION AND FUTURE PERSPECTIVES

The data collected in this review show that sterols, xanthones, flavonoids, phenols, alkaloids, benzoic acid derivatives, terpenes, and glycoside sesquiterpenoids are the predominant constituents in the *Securidaca* genus,

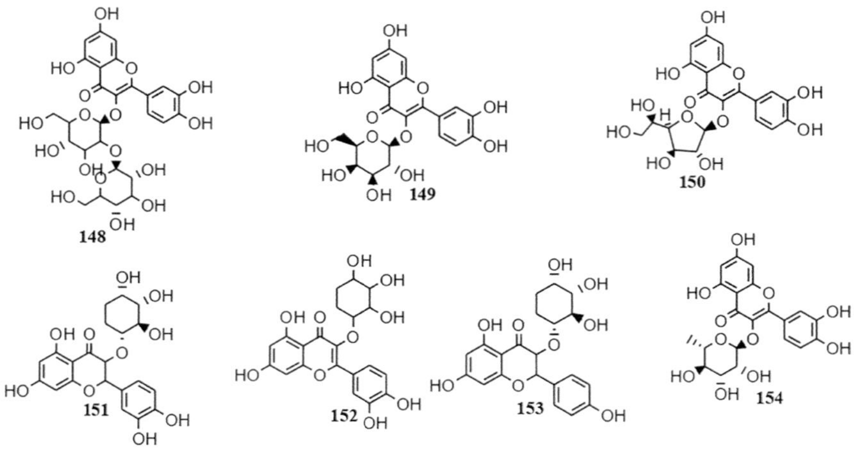

Figure 2.6 Structures of compounds isolated from *S. diversifolia.*

as they have been reported in four of the over 80 species. Despite the presence of these compounds and their biological activities, *Securidaca* species have not been fully studied; thus, subsequent screening is required for other species to open up new avenues for isolating new compounds with new medicinal properties. This review only focused on four of the 80 *Securidaca* species, *S. longepedunculata, S. inappendiculata, S. welwitschii,* and *S. diversifolia,* which have been phytochemically reported in the literature.

Records of *Securidaca* species being utilised to treat different ailments in China, Nigeria, Burkina Faso, and Kenya suggest that these species are valuable ethnomedicine sources. Xanthones are the major compounds isolated from *Securidaca.* Their *in vitro* and *in vivo* bioactivities exhibit substantial anti-inflammatory properties and therapeutic efficacy in inflammatory diseases. This suggests that the genus *Securidaca* is a great source of molecules with medicinal relevance and confirms its traditional applications.

Several bioactivities have been reported for the crude extracts; however, the compounds related to these bioactivities have not been investigated. Most of the isolated compounds have been obtained from the leaves, roots, stems, and barks, despite reports on the traditional medicinal use of various parts (flowers and seeds) of the plants in this species (Keshebo and Choudhury 2015; Bahta 2017). In this regard, there is still much to be done on the phytochemistry and bioactivity of *Securidaca* species. For instance, it is still unknown whether there are notable phytochemical variations across the over 80 species.

REFERENCES

Abubakar, U. S, U. H Danmalam, H. Ibrahim, and B. B. Maiha. 2019. "A Review on African Violet Tree (*Securidaca longipedunculata*): A Traditional Drug with Multiple Medicinal Uses" Specialty Journal of Chemistry 4 (3): 7–14. www.sciarena.com.

Abubakar, U. S., U. H. Danmalam, H. Ibrahim, B. B. Maiha, R. J. Hadiza, and M.S. Abdullahi. 2022a. "Phytochemical Investigation of the Stem Bark of *Securidaca longipedunculata* Fresen (Polygalaceae) (I)." *Journal of Current Biomedical Research* 2 (1): 38–49. https://doi.org/10.54117/jcbr.v2i1.6.

Abubakar, U S, U H Danmalam, H Ibrahim, B B Maiha, R J Hadiza, and M S Abdullahi. 2022b. "LCMS / MS Analysis of the Ethylacetate Extract of *Securidaca longipedunculata* Fresen (Polygalaceae) Stem Bark," *STM Journals* 9 (1): 37–47.

Aderbauer, Birgit, Peter Henning Clausen, Olivia Kershaw, and Matthias F. Melzig. 2008. "In Vitro and in Vivo Trypanocidal Effect of Lipophilic Extracts of Medicinal Plants from Mali and Burkina Faso." *Journal of Ethnopharmacology* 119 (2): 225–31. https://doi.org/10.1016/j.jep.2008.06.024.

Atawodi, S. E., T. Bulus, S. Ibrahim, D. A. Ameh, A. J. Nok, M. Mamman, and M. Galadima. 2003. "*In Vitro* Trypanocidal Effect of Methanolic Extract of Some Nigerian Savannah Plants." *African Journal of Biotechnology* 2 (9): 360–69.

Aymard-Corredor, Gerardo A. 2022. "Novelties in *Securidaca* (*Polygalaceae, Polygaleae*) for the Floras of Colombia and Venezuela." *Harvard Papers in Botany* 27 (2): 143–55. https://doi.org/10.3100/hpib.v27iss2.2022.n2.

Bahta, T.L. 2017. "Essential Oil and Biodiversity Study of Securidaca longipedunculata," *Science and Technology*, 13(May): 91.

Bansal, Poonam, Sunayna Choudhary, Tanvi Taneja, Sonali Sangwan, Bhupesh Gupta, Soniya Goyal, Raman Kumar, and Pooja Sharma. 2019. "Plants Secondary Metabolites: The Key Drivers of the Pharmacological Actions of Medicinal Plants." In *Intech*, 15. https://doi.org/10.5772/intechopen.76139 25.

Bojase, Gomotsang, and Japheth O. Ombito. 2024. "Xanthones from *Securidaca inappendiculata* and *Securidaca longipedunculata*: Chemistry and Biological Activity." *Natural Products Journal* 14 (1): https://doi.org/10.2174/221031 5513666230510123728.

Bruschi, Piero, Michela Morganti, Matteo Mancini, and Maria Adele Signorini. 2011. "Traditional Healers and Laypeople: A Qualitative and Quantitative Approach to Local Knowledge on Medicinal Plants in Muda (Mozambique)." *Journal of Ethnopharmacology* 138 (2): 543–63. https://doi.org/10.1016/j.jep.2011.09.055.

Das, Kuntal. 2009. "Medicinal Plants for Snake Bite Treatment-Future Focus." *Ethnobotanical Leaflets* 13 (July): 508–29.

Debella, Asfaw, Olaf Kunert, Martin G. Schmid, Günter Michl, Franz Bucar, Dawit Abebe, and Ernst Haslinger. 2000. "A Diterpene, a Flavonol Glycoside, and a Phytosterol Glycoside from *Securidaca longipedunculata* and *Entada Abyssinica*." *Monatshefte Fur Chemie* 131 (4): 401–8. https://doi.org/10.1007/s007060050321.

Dibwe, Dya Fita, Suresh Awale, Shigetoshi Kadota, Hiroyuki Morita, and Yasuhiro Tezuka. 2013. "Heptaoxygenated Xanthones as Anti-Austerity Agents from *Securidaca longipedunculata*." *Bioorganic and Medicinal Chemistry* 21 (24): 7663–68. https://doi.org/10.1016/j.bmc.2013.10.027.

Egga, S. E., M. M. Ekwenchi, A. Theophilus, C. Achuenu and S. D. Lokta. 2023. "GC-MS Analysis and the Establishment of Mass Spectrometric Fingerprints of Chemical Constituents Present in Ethanol Extract of *Securidaca longipedunculat*a Leaves." *World Journal of Pharmaceutical Research* 12 (4): 81–94.

Hamburger, M., Gupta, M., and Hostettmann, K. 1985. "Flavonol Glycosides from *Securidaca diversifolia*." *Phytochemistry* 24 (11): 2689–92. https://doi.org/10.1016/S0031-9422(00)80695-0.

Hashim, Ghyda Murad, Mehdi Shahgolzari, Kathleen Hefferon, Afagh Yavari, and Srividhya Venkataraman. 2025. "Plant-Derived Anticancer Therapeutics and Biopharmaceuticals." *Bioengineering* 12 (1): 1–56. https://doi.org/10.3390/bioengineering12010007.

Jayasekara, T. K., P. C. Stevenson, S. R. Belmain, D. I. Farman, and D. R. Hall. 2002. "Identification of Methyl Salicylate as the Principal Volatile Component in the Methanol Extract of Root Bark of *Securidaca longipedunculata* Fers." *Journal of Mass Spectrometry* 37 (6): 577–80. https://doi.org/10.1002/jms.314.

Jayeoye, Titilope John, Fredrick Nwude Eze, Oladipupo Odunayo Olatunde, Sudarshan Singh, Jian Zuo, and Opeyemi Joshua Olatunji. 2021. "Multifarious Biological Applications and Toxic Hg2+ Sensing Potentiality of Biogenic Silver Nanoparticles Based on *Securidaca inappendiculata* Hassk Stem Extract." *International Journal of Nanomedicine* 16 (September): 7557–74. https://doi.org/10.2147/IJN.S325996.

Ji, Cong lan, Sheng Dai, Hong Liu, Ji yang Dong, Chun sheng Liu, and Jian Zuo. 2021. "Polyphenols from *Securidaca inappendiculata* Alleviated Acute Lung Injury in Rats by Inhibiting Oxidative Stress Sensitive Pathways." *Chinese Herbal Medicines* 13 (3): 381–88. https://doi.org/10.1016/j.chmed.2020.09.007.

Ji, Junyang, Qiwen Wang, Maolin Wang, Jianwei Chen, and Xiang Li. 2019. "Chemical Constituents from the Stems of *Securidaca inappendiculata* Hassk." *Fitoterapia* 137 (138): 104271. https://doi.org/10.1016/j.fitote.2019.104271.

Jiang, Hui, Cong Lan Ji, Kui Yang, Wen Zhang, and Jian Zuo. 2018. "Fatty Oil from *Securidaca inappendiculata* Exerted Therapeutic Effects on Adjuvant-Induced Arthritis in Mice via Suppression on Fibroblast-like Synoviocyte." *Kaohsiung Journal of Medical Sciences* 34 (11): 616–25. https://doi.org/10.1016/j.kjms.2018.06.007.

Joseph, Cosam C., M. J. Moshi, J. Sempombe, and M. H.H. Nkunya. 2006. "(4-Methoxy-Benzo[1,3]Dioxol-5-Yl)-Phenylmethanone: An Antibacterial Benzophenone from *Securidaca longipedunculata*." *African Journal of Traditional, Complementary, and Alternative Medicines* 3 (3): 80–86. https://doi.org/10.4314/ajtcam.v3i3.31169.

Kamanzi Atindehou, K., C. Schmid, R. Brun, M. W. Koné, and D. Traore. 2004. "Anti-trypanosomal and Antiplasmodial Activity of Medicinal Plants from Côte d'Ivoire." *Journal of Ethnopharmacology* 90 (2–3): 221–27. https://doi.org/10.1016/j.jep.2003.09.032.

Kang, W. Y., and X. J. Xu. 2008. "Structure of a New Xanthone from *Securidaca inappendiculata*." *Chemistry of Natural Compounds* 44 (4): 432–34. https://doi.org/10.1007/s10600-008-9089-9.

Kang, Wen Yi, Zhi Qiang Ji, and Jin Mei Wang. 2009. "A New Xanthone from the Roots of *Securidaca inappendiculata*." *Chemical Papers* 63 (1): 102–4. https://doi.org/10.2478/s11696-008-0087-y.

Kang, Wen Yi, Zhe Ming Wang, Zheng Quan Li, and Xiao Jie Xu. 2005. "Three New Compounds from *Securidaca inappendiculata*." *Helvetica Chimica Acta* 88 (10): 2771–76. https://doi.org/10.1002/hlca.200590217.

Kaniaru, GeVrald K, Collins M Wafula, Jonathan Sila, and Cecilia Alindalwanyi. 2021. "Community-Based Conservation of the Locally Threatened and Medicinal Securidaca longipedunculata Fresen (Violet Plant); for the People, by the People in the Drylands of Kenya," March: 1–94.

Kerebba, N., A. O. Oyedeji, R. Byamukama and Kuria. 2019. "Pesticidal Activity of Tithonia diversifolia (Hemsl.) A. Gray and Tephrosia Vogelii (Hook f.); Phytochemical Isolation and Characterization: A Review." *South African Journal of Botany* 121:366–76. https://doi.org/10.1016/j.sajb.2018.11.024.

Keshebo, D.L. and M.K Choudhury. 2015. "Phytochemical Investigation of *Securidaca longipedunculata* (*Polygalaceae*) and Structure Elucidation of Benzyl 2-Hydroxy-5-Methoxy Benzoate." *International Journal of Current Microbiology and Applied Sciences* 4 (1): 690–95.

Lannang, A. M., Lontsi, D. N. Ngounou, B. L. Sondengam, A. E. Nkengfack, F. R. van Heerden, and J. C.N. Assob. 2006. "Securidacaxanthone A, a Heptaoxygenated Xanthone from *Securidaca longipedunculata*." *Fitoterapia* 77 (3): 199–202. https://doi.org/10.1016/j.fitote.2006.01.006.

Lognay, G., M. Marlier, D. Seck, and É. Haubruge. 2000. "The Occurrence of 2-Hydroxy-6-Methoxybenzoic Acid Methyl Ester in *Securidaca longipedunculata* Fresen Root Bark." *Biotechnology, Agronomy, Society and Environment* 4 (2): 2–6.

Mahmood, N., P. S. Moore, N. DeTornmasl, F. De Simone, S. Colman, A. J. Hay, and C. Pizza. 1993. "Inhibition of HIV Infection by Caffeoylquinic Acid Derivatives." *Antiviral Chemistry & Chemotherapy* 4 (4): 235–40. https://doi.org/10.1177/095632029300400406.

Meyer, J. J.Marion, N. C. Rakuambo and A. A. Hussein. 2008. "Novel Xanthones from *Securidaca longipedunculata* with Activity against Erectile Dysfunction." *Journal of Ethnopharmacology* 119 (3): 599–603. https://doi.org/10.1016/j.jep.2008.06.018.

Mitaine-Offer, Anne Claire, Nicolas Pénez, Tomofumi Miyamoto, Clément Delaude, Jean François Mirjolet, Olivier Duchamp, and Marie Aleth Lacaille-Dubois. 2010. "Acylated Triterpene Saponins from the Roots of *Securidaca longipedunculata*." *Phytochemistry* 71 (1): 90–94. https://doi.org/10.1016/j.phytochem.2009.09.022.

Mongalo, N. I., L. J. McGaw, J. F. Finnie, and J. Van Staden. 2015. "*Securidaca longipedunculata* Fresen (Polygalaceae): A Review of Its Ethnomedicinal Uses, Phytochemistry, Pharmacological Properties and Toxicology." *Journal of Ethnopharmacology* 165 (February 2015): 215–26. https://doi.org/10.1016/j.jep.2015.02.041.

Moussavi, Nastaran, Wietske van der Ent, Drissa Diallo, Rokia Sanogo, Karl E. Malterud, Camila V. Esguerra and Helle Wangensteen. 2024. "Inhibition of Seizure-Like Paroxysms and Toxicity Effects of *Securidaca longipedunculata* Extracts and Constituents in Zebrafish Danio Rerio." *ACS Chem Neurosci* 15 (3): 617–28. https://doi.org/10.1021/acschemneuro.3c00642.

Obasi, Titus Chukwuemeka, Cornelia Braicu, Bogdan Cezar Iacob, Ede Bodoki, Ancuta Jurj, Lajos Raduly, Ilioara Oniga, Ioana Berindan-Neagoe, and Radu Oprean. 2018. "Securidaca–Saponins Are Natural Inhibitors of AKT, MCL-1, and BCL2L1 in Cervical Cancer Cells." *Cancer Management and Research* 10:5709–24. https://doi.org/10.2147/CMAR.S163328.

Ochora, Douglas O., Esezah Kakudidi, Jane Namukobe, Matthias Heydenreich, Paolo Coghi, Li Jun Yang, Edwin W. Mwakio, et al. 2021. "A New Benzophenone, and the Antiplasmodial Activities of the Constituents of *Securidaca longipedunculata* Fresen (Polygalaceae)." *Natural Product Research* 36 (11): 2758–66. https://doi.org/10.1080/14786419.2021.1925272.

Ochora, Douglas O., Esezah K. Kakudidi, Jane Namukobe, Perpetua Ipulet, Dancan M. Wakoli, Winnie Okore, Edwin W. Mwakio, et al. 2022. "Synergism in Antiplasmodial Activities of Artemether and Lumefantrine in Combination with *Securidaca longipedunculata* Fresen (Polygalaceae)." *Plants* 11 (1). https://doi.org/10.3390/plants11010047.

Ojewole, J. A.O. 2008. "Analgesic, Anti-Inflammatory and Hypoglycaemic Effects of *Securidacal longipedunculata* (Fresen) [*Polygalaceae*] Root-Bark Aqueous Extract." *Inflammopharmacology* 16 (4): 174–81. https://doi.org/10.1007/s10787-007-0016-7.

Olatunji, Opeyemi Joshua, Jian Zuo, and Oladipupo Odunayo Olatunde. 2021. "*Securidaca inappendiculata* Stem Extract Confers Robust Antioxidant and Antidiabetic Effects against High Fructose/Streptozotocin Induced Type 2 Diabetes in Rats. Exploration of Bioactive Compounds Using UHPLC-ESI-QTOF-MS." *Archives of Physiology and Biochemistry* 129 (6): 1187–99. https://doi.org/10.1080/13813455.2021.1921811.

Onobruche, B., M. Ogbunegbe, and O. W. Lamuel. 2017. "Effect of the Root-Bark Extract of *Securidaca longipedunculata* in the Improvement of Fertility in Buck Rabbits." *Advanced Journal of Seed Science and Technology* 4 (5): 95–101.

Osunga, Salome, Omari Amuka, Alex K Machocho, Albert Getabu, and Martin O Onani. 2023. "Some Members of Genus *Cassia* (*Senna*): Their Ethnobotany, Potency and Prospects of Drug Discovery." *Indian Journal of Pharmacy and Drug Studies* 4 (2): 33–57.

Page, Matthew J, Joanne E McKenzie, Patrick M Bossuyt, Isabelle Boutron, Tammy C Hoffmann, Cynthia D Mulrow, Larissa Shamseer, et al. 2021. "The PRISMA 2020 Statement: An Updated Guideline for Reporting Systematic Reviews." *BMJ* 372 (March): n71. https://doi.org/10.1136/bmj.n71.

Pang, Xiaojun, Emmanuel Ayobami Makinde, Fredrick Nwude Eze, and Opeyemi Joshua Olatunji. 2021. "*Securidaca inappendiculata* Polyphenol Rich Extract Counteracts Cognitive Deficits, Neuropathy, Neuroinflammation and Oxidative Stress in Diabetic Encephalopathic Rats via P38 MAPK/Nrf2/HO-1 Pathways." *Frontiers in Pharmacology* 12 (October): 1–16. https://doi.org/10.3389/fphar.2021.737764.

Rakuambo, N. C., J. J. M. Meyer, and A. Hussein. 2004. "Xanthone Isolated from *Securidaca longipendunculata* with Activity against Erectile Dysfunction." *Fitoterapia* 75 (5): 497–99. https://doi.org/10.1016/j.fitote.2004.03.010.

Saha, Aloke, Susmita Moitra, and Tanmay Sanyal. 2022. "Anticancer and Antidiabetic Potential of Phytochemicals Derived from *Catharanthus roseus*: A Key Emphasis to Vinca Alkaloids." *The Basic Handbook of Indian Ethnobotany and Traditional Medicine*, no. October: 1–19. https://doi.org/10.52756/bhietm.2022.e01.001.

Saliu, Jamiu A., and Ayodeji A. Olabiyi. 2017. "Aqueous Extract of *Securidaca longipendunculata* Oliv. and Olax Subscropioidea Inhibits Key Enzymes (Acetylcholinesterase and Butyrylcholinesterase) Linked with Alzheimer's Disease in Vitro." *Pharmaceutical Biology* 55 (1): 252–57. https://doi.org/10.1080/13880209.2016.1258426.

Smith Jr, C. R., R. V. Madrigal, and R. D. Plattner. 1979. "New Conjugated Hydroxydienoic Fatty Acids And." *Biochimica et Biophysics Acta* 572: 314–24.

Stevenson, Philip C., Thamara K. Dayarathna, and Steven R. Belmain, and Nigel C. Veitch. 2009. "Bisdesmosidic Saponins from *Securidaca longipedunculata* Roots: Evaluation of Deterrency and Toxicity to Coleopteran Storage Pests." *Journal of Agricultural and Food Chemistry* 57 (19): 8860–67. https://doi.org/10.1021/jf901599j.

Tikisa, Tahir, Dele Abdissa, and Negera Abdissa. 2019. "Chemical Constituents of *Securidaca longipedunculata* Root Bark and Evaluation of Their Antibacterial Activities." *Ethiopian Journal of Education and Sciences* 14 (2): 1–8. www.ajol.info/index.php/ejesc/article/view/188451.

Timité, Gaoussou, Anne Claire Mitaine-Offer, Tomofumi Miyamoto, Chiaki Tanaka, Thomas Paululat, Clément Delaude, and Marie Aleth Lacaille-Dubois. 2010. "Presenegenin Glycosides from *Securidaca welwitschii*." *Helvetica Chimica Acta* 93 (11): 2237–44. https://doi.org/10.1002/hlca.201000073.

Tommasi N., Piacente S., De Suisone F., Pizza C. De. 1993. "New Sucrose Derivatives from the Bark of *Securidaca longipenduculata*." *Journal of Natural Products* 50 (January): 134–37.

Wedajo, Feyisa, Abera Gure, Melkau Meshesha, Kassim Kedir, Marcel Frese, Norbert Sewald, and Negera Abdissa. 2022. "Cytotoxic Compounds From the Root Bark of *Securidaca longipedunculata*." *Bulletin of the Chemical Society of Ethiopia* 36 (2): 417–22. https://doi.org/10.4314/bcse.v36i2.14.

Yang, Xue Dong, Ning An, Li Zhen Xu, and Yang Shi-Lin. 2002. "New Xanthone Glycosides from *Securidaca inappendiculata*." *Journal of Asian Natural Products Research* 4 (2): 141–45. https://doi.org/10.1080/10286020290027434.

Yang, Xue Dong, Li Zhen Xu, and Shi Lin Yang. 2001. "Xanthones from the Stems of *Securidaca inappendiculata*." *Phytochemistry* 58 (3): 1245–49.

Zha, Haiyan, Zhi Wang, Xuedong Yang, Daqing Jin, Licui Hu, Wenfeng Zheng, Lizhen Xu, and Shilin Yang. 2015. "New Acylated Triterpene Saponins from the Roots of *Securidaca inappendiculata* Hassk." *Phytochemistry Letters* 13: 108–13. https://doi.org/10.1016/j.phytol.2015.05.022.

Zhang, Li jie, Xue Dong Yang, Li Zhen Xu, and Shi lin Yang. 2005. "Three New Xanthones from the Roots of *Securidaca inappendiculata*." *Heterocycles* 65 (7): 1685–90. https://doi.org/10.3987/COM-05-10392.

Zhang, Li Jie, Xue Dong Yang, Li Zhen Xu, Zhong Mei Zou, and Shi Lin Yang. 2005. "A New Sterol Glycoside from *Securidaca inappendiculata*." *Journal of Asian Natural Products Research* 7 (4): 649–53. https://doi.org/10.1080/1028602032000169569.

Zhou, Bin, Yan Wu, Lishe Gan, Seema Dalal, Maria B. Cassera, and Jianmin Yue. 2020. "Structurally Interesting Diarymethane Derivatives from *Securidaca inappendiculata*." *Chinese Journal of Chemistry* 38 (8): 812–16. https://doi.org/10.1002/cjoc.202000094.

Zuo, Jian, Kun-Jun Mao, Fei Yuan, Xiang Li, and Jian Wei Chen. 2014. "Xanthones with Anti-Tumor Activity Isolated from *Securidaca inappendiculata*." *Medicinal Chemistry Research* 23 (11): 4865–71. https://doi.org/10.1007/s00044-014-1051-8.

Zuo, Jian, Yan Xia, Kun Jun Mao, Xiang Li, and Jian Wei Chen. 2014. "Xanthone-Rich Dichloromethane Fraction of *Securidaca inappendiculata*, the Possible Antirheumatic Material Base with Anti-Inflammatory, Analgesic, and Immunodepressive Effects." *Pharmaceutical Biology* 52 (11): 1367–73. https://doi.org/10.3109/13880209.2014.892143.

Chapter 3

Toxicity of plants from the genus *Securidaca* L. (Polygalaceae)

Douglas Ongeri Ochora[1] *and Anywar Godwin U*[2]
[1]Department of Biological Sciences, School of Pure and Applied Sciences, Kisii University, Kisii, Kenya
[2]Department of Plant Sciences, Microbiology and Biotechnology, College of Natural Sciences, Makerere University, Kampala, Uganda
Ethnopharmacology and Zoopharmacognosy, Bernhard Nocht Institute for Tropical Medicine, Hamburg, Germany

3.1 INTRODUCTION

Since time immemorial, human societies have consistently utilized plants for diverse purposes, including medicine, reflecting a long-standing interdependence between humans and the plants (Boukandou et al., 2015). Humanity has continued to use plants for the treatment of various ailments to date. Furthermore, medicinal plants have gained popularity due to their affordability, accessibility, cultural acceptability, and lack of side effects (Mensah et al., 2019). The therapeutic virtues of herbal medication have also contributed to its acceptance. This has resulted in about 80% of the population, especially in the developing world, using medicinal plants for the treatment of various diseases (Jităreanu et al., 2023).

The increased demand for alternative and complementary medicinal products in the market has led to the increased consumption of herbal products. This has necessitated the assessment of the efficacy and toxicity to protect the health of the users (Jităreanu et al., 2023). Addressing these safety concerns is likely to protect humans from exposure to the potentially harmful toxic and carcinogenic effects of herbal medicine (Guo & Mei, 2016).

Toxicity is a major issue in the regulation and legislation of herbal medications. In the United States, the safety of herbs and botanicals as dietary supplements is regulated by the Food and Drug Administration. These herbal products may contain valuable pharmaceuticals, sometimes with significant toxicity. To ensure safety, adverse events are reported, and consumers are informed about the beneficial pharmaceuticals present or nothing of value, side effects, and the potential toxicity (Charen & Harbord, 2020). In other regions of the world, especially in developing

Correspondence: ochoraongeri@gmail.com

DOI: 10.1201/9781003544463-3

countries in Africa, the pharmaceutical bodies of some countries are currently strengthening their legislative and regulatory framework to legalize the use of traditional herbal medicines. Some African countries, including Ghana, South Africa, Uganda, Tanzania, and Ethiopia, now have regulations on the use of herbal medicines (Dubale et al., 2025). There is a need for continued policy development, training initiatives, collaboration among traditional healers and the scientific communitiet, and, finally, the successful reinforcement of the regulatory systems within regions where herbal products are readily accepted (Dubale et al., 2025).

Various factors have contributed to the reported toxicity of herbal medicine. They include variation in the preparation of herbal products among different providers/industries, misidentification of plants, use of wrong plant parts (roots, stems, leaves, stem bark, seeds, flowers, and oils), adulteration when processing herbal products (Charen & Harbord, 2020), inherent toxicity in plants (poisonous plants), environmental pollutants (heavy metals and residues from pesticides), and microbial contamination (toxigenic fungi) (Jităreanu et al., 2023). Other factors that are likely to contribute to plant toxicity include the type and number of secondary metabolites produced by the plants, genetic composition of the plant, climate, soil factors, individual body chemistry, and time of exposure to the herbal medicine (Boukandou et al., 2015). It is therefore imperative that these factors are well understood and mechanisms to avoid or reduce them are implemented. This is likely to reduce the toxicity associated with the use of herbal products.

Plants from the genus *Securidaca* have been used for the treatment of various diseases in traditional herbal medicine (THM). In Africa, *Securidaca longepedunculata* Fresen is widely used to treat various diseases in THM (Mongalo et al., 2015a). In Nigeria, it is considered "the mother of all medicines" (Ngulde et al., 2019). Similarly, *Securidaca inappendiculata* Hassk. has been used for treatment of various diseases in THM in Asian countries, especially in China, for rheumatoid arthritis (Tao et al., 2020; Wang, Li, et al., 2021). Considering the continued and widespread use of plants from the genus *Securidaca* in herbal medicine, this chapter assesses the literature on their toxicity and safety.

3.2 MATERIALS AND METHODS

A scientific literature review was conducted on the PubMed and Google Scholar databases. Subject headings were used as keywords: "toxicity in plants", "toxicity of *Securidaca*", and "toxicity of herbal medicine". Articles and abstracts written in English were the only ones included. Year of publication and affiliation were not considered for exclusion or inclusion.

3.3 RESULTS

3.3.1 Toxicity of herbal medicines

Plants are a promising source for drug discovery and development (Okaiyeto & Oguntibeju, 2021). However, certain plants used in herbal medicine have shown potential toxic effects (Boukandou Mounanga et al., 2015). On the contracty, a study showed that 80% (for the treatment of pain and inflammation in Southern African THM) of the plants tested were considered to be safe (Khumalo et al., 2023). These reports, coupled with the increased popularity of herbal medicine, have created increased attention on the toxicity of herbal medicine (Zhu et al., 2024). On this basis, studies have reported the toxicity of medicinal plants that are used in various cultures, such as Chinese herbal medicine (Liu et al., 2020; Zhu et al., 2024), African THM (Mensah et al., 2019; Okaiyeto & Oguntibeju, 2021), and sub-Saharan African THM (Boukandou et al., 2015).

3.3.2 Methods used to determine the toxicity of herbal medicines

The use of medicinal plants is primarily based on ethnobotanical information that is passed from generation to generation. While the use of ethnobotanical information is preferred, it is challenged by limited validation and minimal toxicity data (Khumalo et al., 2023). In most studies, traditional *in vitro* and *in vivo* methods for the determination of toxicity for herbal medicine have been used. More recent methods, such as omics (metabolomics) and DNA-based methods, have been developed for molecular identification of herbal products. These methods ensure accurate authenticity of herbal medicines, which is key to standardization for safety and toxicity profiling of herbal medicinal products (Jităreanu et al., 2023; Wu & Shaw, 2022).

Metabolomics is considered an important tool because it has been used to determine the quality and safety of medicinal plants. It uses organic compounds with low molecular mass (metabolites), which can discriminate between samples and therefore determine the authenticity of herbal products (Jităreanu et al., 2023; Putri et al., 2013). Likewise, DNA-based methods such as DNA sequencing have been used to detect contaminants and adulteration (Jităreanu et al., 2023). DNA barcoding is also considered a powerful tool since it accurately identifies plant species independent of environmental conditions and morphological limitations that make traditional taxonomic methods unreliable (Wang, Li, et al., 2021). Furthermore, DNA barcoding tools such as PCR-dependent DNA metabarcoding and PCR-free genome skimming/shotgun metagenomics have led to a rapid and authentic identification of poisonous plants (Wang, Zhao, et al., 2021; Wu & Shaw, 2022). These tools offer a potential alternative to circumvent factors that contribute to the toxicity of herbal medicine.

3.3.3 Taxonomy of the genus *Securidaca*

Incorrect identification of plants and using the wrong plant parts have contributed to the reported toxicity of herbal medication (Charen & Harbord, 2020). Therefore, ensuring the correct identification of the plant and plant part is essential to minimize the chances of toxicity associated with herbal medicine.

According to the Plants of the World Online (POWO) 2025, Royal Botanical Gardens, Kew, the genus *Securidaca* was established in 1754 by Linnaeus. *Securidaca* L. belongs to the family Polygalaceae, Hoffmanns. & Link, which has 30 accepted genera. The POWO classifies the genus *Securidaca* as follows: **Kingdom**-Plantae, **Phylum**-Streptophyta, **Class**-Equisetopsida, **Subclass**-Magnoliidae, **Order**-Fabales, **Family**-Polygalaceae, **Genus**-*Securidaca* (POWO, 2025). Based on its chemotaxonomy, the genus *Securidaca* is likely to signify a chemotaxonomic marker for the Polygalaceae family (Lacaille-Dubois & Mitaine-Offer, 2005). The genus *Securidaca* has 60 accepted species that are distributed across various countries, especially in Africa, South America, and Asia. Some specific species of the genus occur in specific regions; most species of the genus occur in South America, while *S. longepedunculata* Fresen and *S. welwitschii* Oliv. occur in Africa, and *S. inappendiculata* Hassk. occurs in Asia, especially in China (POWO, 2025).

Most studies have focused on two species: *S. inappendiculata* and *S. longepedunculata*. Three varieties of *S. longepedunculata* (called violet tree in English) include *S. longepedunculata* var. *longepedunculata*, *S. longepedunculata* var. *parvifolia* Oliv. (Johnson, 1987), and *S. longepedunculata* var. *angustifolia* Robyns. Similarly, two subspecies are recognized for *S. inappendiculata*: *S. inappendiculata* subsp. *corymbosa* (Turcz.) Meijden and *S. inappendiculata* subsp. *inappendiculata* (POWO, 2025). Studies have reported various medicinal properties of the two species. Consequently, toxicity studies of the genus have focused on the same two species.

Various articles have been published on the violet tree with different scientific names; the species name is spelled differently. However, *S. longipedunculata* and *S. longepedunculata* are the most widely used (Moussavi et al., 2024). Both POWO and World Flora Online (WFO) use *S. longepedunculata* (POWO, 2025; WFO, 2025), the International Plant Names Index (IPNI) also uses *S. longepedunculata* with links to both POWO and WFO (IPNI, 2025). The Global Biodiversity Information Facility (GBIF) uses both *S. longipedunculata* and *S. longepedunculata* (GBIF, 2025). Since *Securidaca longepedunculata* appears on the holotype as given by the author (collected from Ethiopia) (GBIF, 2025), it should be the correct spelling of the plant.

Most plants in the genus *Securidaca* grow as shrubs, lianas, or small trees. The main distinctive features of the genus include showy flowers with striking colours ranging from purple to pink, sweetly scented, with eight stamens and filaments that are fused at the base in bouquets on a peduncle (Moussavi et al., 2024). The fruits are typically capsules, some with a membranous wing, and the seeds inside the capsule are often flat or winged (Carter & Ravi, 2024; Mongalo et al., 2015). The stem bark of the plant is pale gray (Figure 3.1) (Moussavi et al., 2024). Leaves have an entire margin, with a slender petiole, arranged alternately with an oblong to narrowly elliptic shape (Figure 3.2).

Figure 3.1 Stem of *S. longepedunculata.*

Figure 3.2 Leaves of *S. longepedunculata.*

3.3.4 Toxicity of *Securidaca longepedunculata*

Mongalo and colleagues have reported on the toxicological studies on *S. longepedunculata* (Mongalo et al., 2015). Histopathological studies of the kidney and liver from Wistar rats treated orally with 300 mg/kg of methanol stem extracts of *S. longepedunculata* displayed no adverse signs of toxicity (Daniel et al., 2022). In the same study, iron oxide nanoparticles from the stem extracts of the plant were considered safe, as the kidneys and liver of Wistar rats displayed normal histoarchitecture, suggesting the absence of acute or chronic damage. Considering the importance of the iron oxide nanoparticles in the treatment of diabetes, the plant has the potential to be used as an antidiabetic with no intrinsic toxic effects (Daniel et al., 2022). In another study, the acute oral toxicity of *S. longepedunculata* at 2,000 mg/kg did not show any toxic signs in BALB/c mice (Kola et al., 2023). The aqueous root extract (400 mg/kg) of *S. longepedunculata* showed a 50% lethal dose (LD_{50}) of 1.74 g/kg and 19.95 mg/kg when tested in albino mice through oral and intraperitoneal routes of administration, respectively (Adeyemi et al., 2010).

The crude extracts and pure compounds from the roots of *S. longepedunculata* have different degrees of cytotoxicity. For instance, the hydro-ethanolic extract and 1,7-dihydroxyxanthone showed the highest activity, with IC_{50} values of 32 µg/mL and 282 µg/mL, respectively, when tested using a free radical method (Hinmara Tina et al., 2025). The DMSO root bark extract of *S. longepedunculata* was considered to be moderately cytotoxic, with a 50% cytotoxic concentration (CC_{50}) value of 98.3 µg/mL, while the ethanol extract exhibited low cytotoxicity (CC_{50} = 90.0 µg/mL) against U87 human glioblastoma cell lines (Anywar et al., 2022). Similarly, the ethanol root extracts inhibited the growth of U87 human glioblastoma cell lines and induced apoptosis in brain tumor cell lines (Ngulde et al., 2019). In the brine shrimp lethality test, the water and organic extracts (methanol and dichloromethane) of *S. longepedunculata* showed moderate cytotoxicity, with an LC_{50} of <1,000 µg/mL (Mokua et al., 2021). In another similar study, the methanol root extracts of *S. longepedunculata* showed a 50% lethal concentration (LC_{50}) of 74.18 µg/mL during a brine shrimp toxicity assay (Adiele et al., 2013). Ethyl alcohol and dichloromethane root extracts of *S. longepedunculata* exhibited high toxicity against zebrafish (*Danio rerio*) larvae, with maximum tolerated concentration (MTC) of 15.7 µg/mL and 11.8 µg/mL, respectively (Moussavi et al., 2024). Benzophenones from roots of *S. longepedunculata* tested at MTC, benzyl-2-hydroxy-6-methoxybenzoate (MTC 12.5 µM), 4,8-dihydroxy-1,2,3,5,6-pentamethoxyxanthone (MTC 25 µM), and 1,7-dihydroxy-4-methoxyxanthone (MTC 6.25 µM), were considered to be the most toxic when tested *in vitro* in larvae (Moussavi et al., 2024). Toxic saponins and methyl salicylate, a suspected nephrotoxin from *S. longepedunculata,* can cause renal damage (Colson & De Broe, 2005). Xanthones from *S. longepedunculata* exhibited potent and selective cytotoxicity against PANC-1 human pancreatic cancer cells (Dibwe et al., 2013).

Despite the widespread use of *S. longepedunculata* for the treatment of various diseases, the plant is likely to cause kidney injury through acute tubular necrosis (Colson & De Broe, 2005). *S. longepedunculata* is considered a toxic plant used for arrow poisoning in Mali, Zambia, Zimbabwe, and Nigeria, and is used as a poison for fishing in Angola and the Ivory Coast (Maiga et al., 2005). This therefore indicated the continued efforts to determine a safe dosage of the plant.

3.3.5 Toxicity of *Securidaca inappendiculata*

S. inappendiculata is a plant used for the treatment of various diseases, especially in Chinese herbal medication. It is the most studied Chinese plant of the genus *Securidaca* (Wang, Li, et al., 2021; Zhang et al., 2023). It is considered an anti-rheumatic herbal medicine native to southern China (Tao et al., 2020). Therefore, various studies have reported the toxicity of this plant. Based on the increasing global popularity of Chinese traditional medicine, there has been a corresponding increase in attention on the toxicity of Chinese herbal medicine in the last decade (2013–2022) (Zhu et al., 2024). Most of these Chinese herbal medicines are associated with hepatotoxicity and nephrotoxicity, with apoptosis considered as the main mechanism of toxicity (Zhu et al., 2024). The reported nephrotoxicity in Chinese herbal medications is mainly caused by flavonoids, alkaloids, glycosides, and aristolochic acid (Charen & Harbord, 2020).

The dichloromethane and ethyl acetate fractions from alcohol stem extract of *S. inappendiculata* were considered safe when tested in ICR mice, while methanol and acetone fractions of the same extract were considered toxic (Zuo et al., 2014). A study on silver nanoparticles from the extract of *S. inappendiculata* stem showed no risk of haemolysis of erythrocytes and is likely to have biocompatibility; therefore, the plant has the potential to create a new environmentally stable stabilizer without adverse effects to human health (Jayeoye et al., 2021). Additionally, phytochemicals such as xanthones from *S. inappendiculata* were considered non-toxic, with LD_{50} values above 100 mg/kg, when tested at a concentration of 100 mg/kg using DBA/1 mice dosed through intravenous injection (Zuo, Xia, Li, et al., 2014). Similarly, triterpene saponins from the roots of *S. inappendiculata* showed cytotoxic activity through cell death-inducing activity displayed on the tumor (macrophage-like) cell lines (Yui et al., 2001). These and continued toxicity studies of the plant will act as a guide on the safe usage of the plant in herbal medicine.

3.4 CONCLUSION

Among various plants of the genus *Securidaca*, only *S. longepedunculata* from Africa and *S. inappendiculata* from China are widely used in herbal medicine and are the most studied plants. Similarly, toxicity studies have focused on the two plants. Most extracts and compounds from the plants are considered safe at given concentrations. However, more studies should be conducted to confirm the toxicity of the plants from the genus *Securidaca*, especially through clinical trials. This will act as a guide to the safe use of these plants. The toxicity of herb–drug and herb–herb combinations of plants from the genus *Securidaca* remains unknown. Similarly, the toxicity levels of other plants of the genus *Securidaca* should be studied since the phytochemistry and therapeutic efficacy of the plants are likely to be the same. This toxicity information will contribute to information toward their legislation for the safe medical benefits of herbal products.

REFERENCES

Adeyemi, O. O., Akindele, A. J., Yemitan, O. K., Aigbe, F. R., & Fagbo, F. I. (2010). Anticonvulsant, anxiolytic and sedative activities of the aqueous root extract of *Securidaca longepedunculata* Fresen. *Journal of Ethnopharmacology*, *130*(2), 191–195. https://doi.org/10.1016/j.jep.2010.04.028

Adiele, R. C., Fakae, B. B., & Isuzu, I. U. (2013). Anthelmintic activity of *Securidaca longepedunculata* (Family: Polygalaceae) root extract in mice, *in vitro* and *in vivo*. *Asian Pacific Journal of Tropical Medicine*. https://doi.org/10.1016/S1995-7645(13)60150-9

Anywar, G. U., Kakudidi, E., Oryem-Origa, H., Schubert, A., & Jassoy, C. (2022). Cytotoxicity of medicinal plant species used by traditional healers in treating people suffering from HIV/AIDS in Uganda. *Frontiers in Toxicology*, *4*(May), 1–8. https://doi.org/10.3389/ftox.2022.832780

Boukandou, M., Mewono, L., & Aboughe Angone, S. (2015). Toxicity studies of medicinal plants used in sub-Saharan Africa. *Journal of Ethnopharmacology*, *174*, 618–627. https://doi.org/10.1016/j.jep.2015.06.005

Carter, P., & Ravi, G. (2024). *Securidaca* L . (Polygalaceae): A new generic record to the flora of Andhra Pradesh, India. *Indian Botanical Society*, 19–21. https://doi.org/10.61289/jibs2024.11.08.152

Charen, E., & Harbord, N. (2020). Toxicity of herbs, vitamins, and supplements. *Advances in Chronic Kidney Disease*, *27*(1), 67–71. https://doi.org/10.1053/j.ackd.2019.08.003

Colson, C. R. D., & De Broe, M. E. (2005). Kidney injury from alternative medicines. *Advances in Chronic Kidney Disease*, *12*(3), 261–275. https://doi.org/10.1016/j.ackd.2005.03.006

Daniel, A. I., Umar, M. B., Tijani, O. J., & Muhammad, R. (2022). Antidiabetic potentials of green-synthesized alpha iron oxide nanoparticles using stem extract of *Securidaca longipedunculata. International Nano Letters*, *12*(3), 281–293. https://doi.org/10.1007/s40089-022-00377-x

Dibwe, D. F., Awale, S., Kadota, S., Morita, H., & Tezuka, Y. (2013). Heptaoxygenated xanthones as anti-austerity agents from *Securidaca longepedunculata. Bioorganic and Medicinal Chemistry, 21*(24), 7663–7668. https://doi.org/10.1016/j.bmc.2013.10.027

Dubale, S., Usure, R. E., Mekasha, Y. T., Hasen, G., Hafiz, F., Kebebe, D., & Suleman, S. (2025). Traditional herbal medicine legislative and regulatory framework: a cross-sectional quantitative study and archival review perspectives. *Frontiers in Pharmacology*, *16*(January), 1–22. https://doi.org/10.3389/fphar.2025.1475297

Global Biodiversity Information Facility (GBIF). (2025). www.gbif.org/search?q=SECURIDACA%20longipedunculata (accessed 10 June 2025).

www.gbif.org/occurrence/3046121603 (accessed 10th June 2025).

Guo, X., & Mei, N. (2016). Aloe vera: A review of toxicity and adverse clinical effects. *Journal of Environmental Science and Health – Part C Environmental Carcinogenesis and Ecotoxicology Reviews*, *34*(2), 77–96. https://doi.org/10.1080/10590501.2016.1166826

Hinmara Tina, J., Tagousop, C. N., Abdouraman, B., Gbaweng, J. A., Feunaing Toko, R., Djithe Djouakoua, J., Lita Morsia, E., Henoumont, C., Laurent, S., & Talla, E. (2025). Securidacaxanthone D: A new xanthone with free radical scavenging activity from the roots of *Securidaca longepedunculata* (Polygalaceae). *Natural Product Research*, *0*(0), 1–7. https://doi.org/10.1080/14786419.2025.2484268

International Plant Names Index (IPNI). (2025). www.ipni.org/search?q=Securidaca%20longepedunculata (accessed 10th June 2025).

Jayeoye, T. J., Eze, F. N., Olatunde, O. O., Singh, S., Zuo, J., & Olatunji, O. J. (2021). Multifarious biological applications and toxic Hg2+ sensing potentiality of biogenic silver nanoparticles based on *Securidaca inappendiculata* hassk stem extract. *International Journal of Nanomedicine*, *16*, 7557–7574. https://doi.org/10.2147/IJN.S325996

Jităreanu, A., Trifan, A., Vieriu, M., Caba, I. C., Mârțu, I., & Agoroaei, L. (2023). Current trends in toxicity assessment of herbal medicines: A narrative review. *Processes*, *11*(1). https://doi.org/10.3390/pr11010083

Johnson, C. T. (1987). Taxonomy of the African species of *Securidaca* (Polygalaceae). *South African Journal of Botany*, *53*(1), 5–11. https://doi.org/10.1016/S0254-6299(16)31465-X

Khumalo, G. P., Van Wyk, B. E., Feng, Y., & Cock, I. E. (2023). Toxicity and phytochemical properties of southern African medicinal plants used traditionally to treat pain and inflammatory ailments. *South African Journal of Botany*, *160*, 102–122. https://doi.org/10.1016/j.sajb.2023.07.005

Kola, P., Manjula, S. N., Metowogo, K., Madhunapantula, S. V., & Eklu-Gadegbeku, K. (2023). Four Togolese plant species exhibiting cytotoxicity and antitumor activities lightning polytherapy approach in cancer treatment. *Heliyon*, *9*(3), e13869. https://doi.org/10.1016/J.HELIYON.2023.E13869

Lacaille-Dubois, M. A., & Mitaine-Offer, A. C. (2005). Triterpene saponins from Polygalaceae. *Phytochemistry Reviews*, *4*(2–3), 139–149. https://doi.org/10.1007/s11101-005-2606-6

Liu, R., Li, X., Huang, N., Fan, M., & Sun, R. (2020). Toxicity of traditional Chinese medicine herbal and mineral products. In *Advances in Pharmacology* (1st ed., Vol. 87). Editors: Mehmet Can Atayik, Ufuk Cakatay. Elsevier Inc. https://doi.org/10.1016/bs.apha.2019.08.001

Maiga, A., Diallo, D., Fane, S., Sanogo, R., Paulsen, B. S., & Cisse, B. (2005). A survey of toxic plants on the market in the district of Bamako, Mali: Traditional knowledge compared with a literature search of modern pharmacology and toxicology. *Journal of Ethnopharmacology*, *96*(1–2), 183–193. https://doi.org/10.1016/j.jep.2004.09.005

Mensah, M., Komlaga, G., D. Forkuo, A., Firempong, C., K. Anning, A., & A. Dickson, R. (2019). Toxicity and safety implications of herbal medicines used in Africa. *Herbal Medicine*. https://doi.org/10.5772/intechopen.72437

Mongalo, N. I., McGaw, L. J., Finnie, J. F., & Staden, J. Van. (2015). *Securidaca longipedunculata* Fresen (Polygalaceae): A review of its ethnomedicinal uses, phytochemistry, pharmacological properties and toxicology. *Journal of Ethnopharmacology*, *165*, 215–226. https://doi.org/10.1016/j.jep.2015.02.041

Mokua, S. K., Mbaria, J. M., Maitho, T. E., & Moriasi, G. A. (2021). Ethnobotanical documentation, phytochemical screening, and cytotoxicity evaluation of medicinal plants used to manage snakebite envenomation in Mwingi West Subcounty, Kenya. *Evidence-Based Complementary and Alternative Medicine*, *2021*. https://doi.org/10.1155/2021/4167296

Mongalo, N. I., McGaw, L. J., Finnie, J. F., & Staden, J. Van. (2015a). *Securidaca longipedunculata* Fresen (Polygalaceae): A review of its ethnomedicinal uses, phytochemistry, pharmacological properties and toxicology. *Journal of Ethnopharmacology*, *165*, 215–226. https://doi.org/10.1016/j.jep.2015.02.041

Moussavi, N., van der Ent, W., Diallo, D., Sanogo, R., Malterud, K. E., Esguerra, C. V., & Wangensteen, H. (2024). Inhibition of seizure-like paroxysms and toxicity effects of *Securidaca longepedunculata* extracts and constituents in zebrafish Danio rerio. *ACS Chemical Neuroscience*, *15*(3), 617–628. https://doi.org/10.1021/acschemneuro.3c00642

Ngulde, S. I., Sandabe, U. K., Abounader, R., Dawson, T. K., Zhang, Y., Iliya, I., & Hussaini, I. M. (2019). Ethanol extract of *Securidaca longipedunculata* induces apoptosis in brain tumor (U87) cells. *BioMed Research International*, *2019*, 1–5. https://doi.org/10.1155/2019/9826590

Okaiyeto, K., & Oguntibeju, O. O. (2021). African herbal medicines: Adverse effects and cytotoxic potentials with different therapeutic applications. *International Journal of Environmental Research and Public Health*, *18*(11). https://doi.org/10.3390/ijerph18115988

Plants of the World Online (POWO). (2025). https://powo.science.kew.org/results?q=Securidaca%20longepedunculata (accessed 10th June 2025).

Putri, S. P., Nakayama, Y., Matsuda, F., Uchikata, T., Kobayashi, S., Matsubara, A., & Fukusaki, E. (2013). Current metabolomics: Practical applications. *Journal of Bioscience and Bioengineering*, *115*(6), 579–589. https://doi.org/10.1016/J.JBIOSC.2012.12.007

Tao, M. qing, Ji, C. lan, Wu, Y. jin, Dong, J. yang, Li, Y., Olatunji, O. J., & Zuo, J. (2020). 1,7-dihydroxy-3,4-dimethoxyxanthone iinhibits lipopolysaccharide-induced inflammation in RAW264.7 macrophages by suppressing TLR4/NF-κB signaling cascades. *Inflammation*, *43*(5), 1821–1831. https://doi.org/10.1007/s10753-020-01256-3

Wang, D. D., Li, Y., Wu, Y. J., Wu, Y. L., Han, J., Olatunji, O. J., Wang, L., & Zuo, J. (2021a). Xanthones from *Securidaca inappendiculata* antagonized the antirheumatic effects of methotrexate *in vivo* by promoting its secretion into urine. *Expert Opinion on Drug Metabolism and Toxicology*, *17*(2), 241–250. https://doi.org/10.1080/17425255.2021.1843634

Wang, J., Zhao, J., Yu, W., Wang, S., Bu, S., Shi, X., & Zhang, X. (2021b). Rapid identification of common poisonous plants in China using DNA Barcodes. *Frontiers in Ecology and Evolution*, *9*(August), 1–10. https://doi.org/10.3389/fevo.2021.698418

World Flora Online (WFO). 2025. https://wfoplantlist.org/taxon/wfo-0000503535-2024-12?page=1 (accessed 10th June 2025).

Wu, H. Y., & Shaw, P. C. (2022). Strategies for molecular authentication of herbal products: from experimental design to data analysis. *Chinese Medicine (United Kingdom)*, *17*(1), 1–15. https://doi.org/10.1186/s13020-022-00590-y

Yui, S., Ubukata, K., Hodono, K., Kitahara, M., Mimaki, Y., Kuroda, M., Sashida, Y., & Yamazaki, M. (2001). Macrophage-oriented cytotoxic activity of novel triterpene saponins extracted from roots of *Securidaca inappendiculata*. *International Immunopharmacology*, *1*(11), 1989–2000. https://doi.org/10.1016/S1567-5769(01)00126-6

Zhang, S. sa, Wu, Y. jin, Pan, S., Wang, Q. hai, Chen, W. gang, Wang, Z. fang, Xu, J. ying, Zuo, J., & Yin, Q. (2023). Xanthones from *Securidaca inappendiculata* antagonizes the anti-rheumatic effect of methotrexate by inhibiting reduced folate carrier 1. *Immunopharmacology and Immunotoxicology*, *45*(1), 16–25. https://doi.org/10.1080/08923973.2022.2103707

Zhu, K. X., Wu, M., Bian, Z. L., Han, S. L., Fang, L. M., Ge, F. F., Wang, X. Z., & Xie, S. F. (2024). Growing attention on the toxicity of Chinese herbal medicine: a bibliometric analysis from 2013 to 2022. *Frontiers in Pharmacology*, *15*(February), 1–11. https://doi.org/10.3389/fphar.2024.1293468

Zuo, J., Xia, Y., Li, X., & Chen, J. W. (2014). Xanthones from *Securidaca inappendiculata* exert significant therapeutic efficacy on adjuvant-induced arthritis in mice. *Inflammation*, *37*(3), 908–916. https://doi.org/10.1007/s10753-014-9810-8

Zuo, J., Xia, Y., Mao, K. J., Li, X., & Chen, J. W. (2014). Xanthone-rich dichloromethane fraction of *Securidaca inappendiculata*, the possible antirheumatic material base with anti-inflammatory, analgesic, and immunodepressive effects. *Pharmaceutical Biology*, *52*(11), 1367–1373. https://doi.org/10.3109/13880209.2014.892143

Chapter 4

Ethnobotany of the genus *Securidaca* L.

Anywar Godwin U[1] *and Douglas Ongeri Ochora*[2]

[1]Department of Plant Sciences, Microbiology and Biotechnology, College of Natural Sciences, Makerere University, Kampala, Uganda
Ethnopharmacology and Zoopharmacognosy, Bernhard Nocht Institute for Tropical Medicine, Hamburg, Germany
[2]Department of Biological Sciences, School of Pure and Applied Sciences, Kisii University, Kisii, Kenya

4.1 INTRODUCTION

Securidaca L., is a pantropical genus of Polygalaceae family (Aymard-Corredor, 2022) with 59 accepted species (Plants of the World Online, 2025). The largest diversity is found in the Neotropics (Aymard et al., 2004); a few species are found in Africa and Southeast Asia (Aymard-Corredor, 2022). Ethnobotanical information about neotropical species of *Securidaca* is not commonly found in the literature. The ethnobotanical literature provided contains information on the two plant species reported in Africa and the other species discussed are outside of Africa (Aymard-Corredor, 2022).

4.1.1 *Securidaca longepedunculata* Fresen

Securidaca longepedunculata is a small tree with alternate leaves which are variable in size, shape, and crowded towards the stem tips (Medicinal Plants of South Africa, 2010). Its flowers are small, pink, or purple in colour, sweet scented, and are usually produced in early summer (Van Wyk et al., 2005). The fruits of this plant are heavily veined, smooth, oblong, purplish green when young, and also possess a membranous wing (Coates Palgrave, 1977).

4.2 LOCAL NAMES

In South African languages, this plant species is called violet tree, fibre tree, or Rhodesian violet (English), krinkhout (Afrikaans), umfufu (Ndebele), mupesu (TshiVenda), and mmaba in both Sotho and Tswana (Mongalo et al., 2015). In other African countries, various names in different cultural and ethnic groups have been used: es a manahi (Amharic); saggat, alali (Arabic); mwinda (Lozi); mutata (Lunda); mwinda, mpuluka (Nyanja); mupapi (Bemba); uwar magunguna, sanya (Hausa); yodo, juto, jodo (Mandinka); mufufu (Shona); muteya, mzigi, Chipvufana, mufufu, munyapunyapu, munyazvirombo, mutangeni, umfufu (Swahili); shotora (Tigrigna); njefu,

 DOI: 10.1201/9781003544463-4

bwazi, mufufuma (Tongan); fouf (Wolof); ipeta (Yoruba) (Orwa et al., 2009). In Uganda, it is known by several names in different languages. These include Elilyoi/elilie (Ateso); Lilo (Luganda); Oiyofe (Lugbara); Wadambasima (Lugishu); Mwiabala (Lugwe); Loloyi (Lugwere); Aliya, lalia, lalon (Luo: Acholi); Lilyo (Luo Japhadhola); Elila (Luo Langi); Mukondwa (Lusoga); Lio (Madi; Mweya (Runyankore); Nkondwe (Runyoro) (Katende et al., 1995), Nyilia (Luo: Alur).

4.3 DISTRIBUTION AND CONSERVATION

S. longepedunculata is widely distributed in tropical and subtropical areas of Africa, which include Angola, Benin, Botswana, Cameroon, Chad, Eritrea, Ethiopia, Gambia, Ghana, Kenya, Malawi, Namibia, Niger, Nigeria, Rwanda, Senegal, South Africa, Sudan, Tanzania, Uganda, and Zambia (Baloyi & Tshisikhawe, 2009).

Although *S. longepedunculata* is protected under provincial and national legislation in many countries, it is still one of the most traded medicinal plants on the African continent (Moeng, 2010). This important medicinal plant is threatened by over-harvesting; this may be due to the fact that the root is the most commonly used part, which makes the plant difficult to survive over-harvesting (Abubakar et al., 2019). Other factors threatening its survival include urbanization, expansion of agricultural activities, and anthropogenic and environmental conditions, such as bush fires and droughts (Oni et al., 2014).

4.4 TRADITIONAL MEDICINAL USES OF *SECURIDACA LONGEPEDUNCULATA*

The roots of *S. longepedunculata* have been used as a remedy to treat more than 24 diseases in Africa (Abubakar et al., 2019). The leaves and roots are used to treat a variety of symptoms in people living with HIV/AIDS, including fatigue, joint and chest pain, cough, ulcers, and oral sores, in Uganda (Anywar et al., 2020a). The roots are also given to boost the immune system of immune-compromised people, often prepared as a decoction or used to make clay tablets (Anywar et al., 2020b).

The root of *S. longepedunculata* is traditionally used to manage fungal infections, fever, malaria, gonorrhoea, headaches, cancer, rheumatism, tuberculosis, diabetes, venereal diseases, syphilis, sexual impotence, pains, epilepsy, convulsions, constipation, pneumonia, backache, blood purification, sexually transmitted infections, skin infections, and also used as an aphrodisiac (Mustapha, 2013b). The leaves of *S. longepedunculata* are locally used to treat epilepsy, headaches, stomach ache, infertility, snakebite, toothache, cancer, skin infections, dislocated jaw, for contraceptive purposes, and to expel the placenta (Mustapha, 2013b), while its stem bark is traditionally used to treat epilepsy, stomach ache, venereal diseases, skin diseases, dysentery, malaria, typhoid, inflammation, chest complaints, constipation, viral infections, snakebites, and infertility problems (Kadiri et al., 2013). The whole plant is used as a mouth wash to treat infections such as oral candidiasis, and other opportunistic infections associated with HIV (Chinsembu & Hedimbi, 2010). It is used as an antidote for snakebite. The bark is used as an abortifacient and a root infusion is used to relieve toothache (Katende et al., 1995; Mustapha, 2013b).

Smoke from the burning roots of *S. longepedunculata*, combined with that of *Zanthoxylum zanthoxyloides*, is inhaled to treat malaria and fever in Uganda (Hamill et al., 2003). A root decoction may also be drunk to treat fever, malaria, hernias, gonorrhoea, palpitations, headaches, oedema, rheumatism, diabetes, sexual impotence, toothache, and fungal infections (Maroyi, 2013). An infusion of the soaked root bark may be drunk as an aphrodisiac or mixed with other medicines and used as an emetic (Mabogo, 2012). Alternatively, a root decoction may be drunk in beer as an aphrodisiac (Motlhanka & Nthoiwa, 2012).

The root bark is pulverised in water and the resulting mixture is inhaled or used to wash the head, treating strong headache (Nordeng et al., 2013). A handful of roots are combined with the roots of *Sphedamnocarpus pruriens* subsp. *pruriens* for treating people believed to be possessed by evil spirits, while the powdered root is mixed with porridge and eaten to treat epilepsy and convulsions (Sobiecki, 2008). The decoction from the root is drunk or applied topically to treat cancer (Ashidi et al., 2010). The roots may also be ground into powder form, dissolved in water, and taken orally for constipation, pneumonia, backache, blood purification, sexually transmitted infections, and as an aphrodisiac (Viol, 2009).

The dried roots are soaked in water, along with *Citrus aurantiifolia* and taken for three days to treat constipation. The dried root is also boiled in water along with that of *Annona senegalensis* and used to treat pneumonia (Mustapha, 2013a). The dried root is also ground into powder, along with that of *Parkia biglobosa,* and then taken with cow's milk as a sexual boost (Mongalo et al., 2015). The pounded root may be mixed with that of *Zanthoxylum humile* and taken with soft porridge to treat erectile dysfunction (Semenya & Potgieter, 2013).

The fresh leaves are made into paste with little or no water along with the bark of *Gardenia erubescens* and applied externally twice a day to treat skin cancer (Mustapha, 2013a). The fresh leaves are made into paste with little or no water along with leaves of *Jussiaea suffruticosa* and shea butter and applied externally twice a day to treat a variety of skin infections (Mongalo et al., 2015). Dry leaves are also ground into powder and put into the fire and the resulting smoke is inhaled to treat headaches, while the boiled leaves are taken orally for contraceptive purposes (Mustapha, 2013a). The leaves are either chewed fresh or orally and nasally administered to treat epilepsy, headaches, stomach ache, infertility, snakebite, toothache, and to expel the placenta (Augustino et al., 2011).

One cup from a whole plant decoction may be taken orally three times a day for three to four days to treat malaria (Nguta et al., 2010). The decoction of the whole plant may either be drunk or used to wash the mouth and treat infections which include oral candidiasis, and other opportunistic infections associated with HIV/AIDS (Chinsembu & Hedimbi, 2010).

A spoonful of powdered stem bark is mixed with *Mondia whitei* (stem bark), *Uvaria afzelii* (root bark), *Allium ascalonicum* (bulb), and *Parkia biglobosa* (seeds) is then taken with hot porridge to treat a variety of viral infections (Borokini et al., 2013). The dried bark is ground into a powder and taken orally with cow's milk or porridge for 14 days to treat dysentery (Mustapha, 2013b). A decoction from the stem bark may be taken orally to treat stomach ache, headaches, inflammation, chest complaints, abortion, jaundice, constipation, snakebites, infertility problems, epilepsy, and venereal diseases (Kadiri et al., 2013). The powdered stem bark is also mixed with hot water and taken orally to treat syphilis and gonorrhoea (Hedimbi & Chinsembu, 2012). It is clear that S. *longependuncul ata* is widely used for treating malaria, sexually transmitted infections, enhancing sexual performance, headaches, toothaches, pain, epilepsy, malariaand cough across the continent.

The combination of the extracts of the roots, stems, and leaves of *Securidaca longependuncul ata* with artemether and lumefantrine exhibited effective antiplasmodial activity against *Plasmodium falciparum* welch strains (Ochora et al., 2021).

The roots of several species in the Polygalaceae family are important sources of methyl salicylate (da Rocha et al., 2012), which is a compound used in traditional medicine, mainly as an anti-inflammatory, analgesic, expectorant, or anti-rheumatic remedy (Effmert et al., 2005). This natural compound is widely used in flavouring for foods, candies, and beverages, and in the pharmaceutical industry for making topical ointments to treat muscle pains (Victório et al., 2011).

4.4.1 *Securidaca welwitschii* Oliv.

This plant species is a liana widely distributed in Guinea, Cameroon, Democratic Republic of Congo, Angola, and Uganda (Davreux & Delaude, 1971). *S. welwitschii* is a large shrub often climbing to a great height (10–25 m) with smooth and glabrous branches (Chuks et al., 2016). This plant contains methyl salicylate and one major saponin deriving from presenegenin (Timité et al., 2010).

4.5 TRADITIONAL MEDICINAL USES OF *SECURIDACA WELWITSCHII*

The use *S. welwitschii* in traditional medicine is rare except for eye infections (Davreux & Delaude, 1971). In Eastern Nigeria, the decoction of the root of *S. welwitschii* is used to deworm people, while in Gabon the stem sap is used for eye trouble (Bytebier, 2008).

4.5.1 *Securidaca inappendiculata* Hassk.

Securidaca inappendiculata Hassk. is a plant native to tropical America and it is widely grown in the tropical regions of China, especially Guangdong, Guangxi, Hainan, and Yunnan provinces (Olajide et al., 1998). The stems of *S. inappendiculata* have been widely employed traditionally in Chinese medicine for many centuries as an agent for promoting blood circulation, removing blood stasis, treating fractures, and rheumatoid arthritis, alleviating pains, and healing wounds (Ji et al., 2019).

Since the biological effects of *S. inappendiculata* are attributed to its bioactive chemical constitutes, some researches have conducted extensive investigations on them. These include xanthones, benzophenones, lignan glycosides, triterpene glycosides, and phenolic glycosides (Wang et al., 2021). In addition, several pharmacological studies have shown that *S. inappendiculata* has beneficial effects against arthritis, cancer, inflammation, pain, and plays a role in α-glucosidase inhibition (Ji et al., 2019).

4.6 OTHER USES OF *SECURIDACA* SPP.

A fibre obtained from the inner bark can be used like cotton to weave a coarse cloth, fishing nets and baskets (Abubakar et al., 2019), bead strings and thread to sew bark cloth (Katende et al., 1995), for hut construction, brooms, and poles (Wyk & Gericke, 2000). It is a beautiful flowering tree with potential as an ornamental in parks and gardens (Katende et al., 1995). The bark and leaves of *Securidaca* spp. (e.g., *S. pyramidalis* Sprague, *S. retusa* Benth) soaked in water produce a very good soap used in the upper Rio Orinoco basin and upper Rio Negro region of Colombia and Venezuela (Aymard et al., 2004).

The roots of "African Violet tree" (*S. longepedunculata*) have been used as a remedy to treat more than 24 diseases in Africa (Abubakar et al., 2019).

4.7 CONCLUSION

Securidaca longepedunculata is by far the most popular and most widely used medicinal plant from the genus *Securidaca* in Africa. This is followed by *S. inappendiculata* in Asia. As such, most of the literature available on the ethnobotanical aspects of members of this genus are from *S. longepedunculata*.

S. longepedunculata thus occupies an important position in the ethnobotanical literature, especially as a medicinal a plant for treating various diseases. It thus plays a significant role as a therapeutic agent, but also plays other important roles in the day-to-day lives of indigenous communities where it grows. The non-medicinal uses include its use for making coarse cloth, fishing nets and baskets, soap, hut construction, brooms, and poles and its role as an ornamental plant. The huge demand and popularity of especially *S. longepedunculata* and the fact that its roots are widely used make it vulnerable to the harvesting pressures and threaten its survival.

REFERENCES

Abubakar, U. S., Danmalam, U. H., Ibrahim, H., & Maiha, B. B. (2019). A review on African violet tree (Securidaca longipedunculata Fresen): A traditional drug with multiple medicinal uses. *Specialty Journal of Chemistry*, *4*(3–2019), 7–14.

Anywar, G., Kakudidi, E., Byamukama, R., Mukonzo, J., Schubert, A., & Oryem-Origa, H. (2020a). Indigenous traditional knowledge of medicinal plants used by herbalists in treating opportunistic infections among people living with HIV/AIDS in Uganda. *Journal of Ethnopharmacology*, *246*. https://doi.org/10.1016/j.jep.2019.112205

Anywar, Godwin, Kakudidi, E., Byamukama, R., Mukonzo, J., Schubert, A., & Oryem-Origa, H. (2020b). Medicinal plants used by traditional medicine practitioners to boost the immune system in people living with HIV/AIDS in Uganda. *European Journal of Integrative Medicine*, *35*, 101011. https://doi.org/10.1016/j.eujim.2019.101011

Ashidi, J. S., Houghton, P. J., Hylands, P. J., & Efferth, T. (2010). Ethnobotanical survey and cytotoxicity testing of plants of South-western Nigeria used to treat cancer, with isolation of cytotoxic constituents from Cajanus cajan Mill sp. leaves. *Journal of Ethnopharmacology*, *128*(2), 501–512. https://doi.org/10.1016/j.jep.2010.01.009

Augustino, S., Hall, J. B., Makonda, F. B. S., & Ishengoma, R. C. (2011). Medicinal resources of the Miombo woodlands of Urumwa, Tanzania: plants and its uses. *Journal of Medicinal Plants Research*, *5*(27), 6352–6372.

Aymard-Corredor, G. A. (2022). Novelties in Securidaca (Polygalaceae, Polygaleae) for the Floras of Colombia and Venezuela. *Harvard Papers in Botany*, *27*(2), 143–155.

Aymard, G. A., Berry, P. E., & Eriksen, B. (2004). Polygalaceae. *Flora of the Venezuelan Guayana*, *8*, 316–347.

Baloyi, O., & Tshisikhawe, M. P. (2009). The population ecology of Securidaca longepedunculata Fresen in Nylsvley Nature Reserve, Limpopo Province, RSA. *South African Journal of Botany*, *75*(2), 430. https://doi.org/10.1016/j.sajb.2009.02.134

Borokini, T. I., Clement, M., Dickson, N. J., & Edagbo, D. E. (2013). Ethnobiological survey of traditional medicine practice for circulatory and nervous system related diseases in Oyo State, Nigeria. *Topclass Journal of Herbal Medicine*, *2*, 111–120.

Bytebier, B. (2008). Flora of Tropical East Africa. *Journal of East African Natural History*, *97*(2), 259–260. https://doi.org/10.2982/0012-8317-97.2.259

Chinsembu, K. C., & Hedimbi, M. (2010). An ethnobotanical survey of plants used to manage HIV/AIDS opportunistic infections in Katima Mulilo, Caprivi region, Namibia. *Journal of Ethnobiology and Ethnomedicine*, *6*(1), 1–9. https://doi.org/10.1186/1746-4269-6-25

Coates Palgrave, K. (1977). Trees of southern Africa. *Veld & Flora*, *63*(4), 8.

da Rocha, J. L. C., Pastore, J. F. B., Brandão, H. N., Azeredo, A., David, J. P., Santos, E. O. dos, & David, J. M. (2012). Quantificação de salicilato de metila em quatro gêneros de Polygalaceae, por CLAE-DAD. *Química Nova*, *35*, 2263–2266.

Davreux, M., & Delaude, C. (1971). Saponins of the Polygalaceae: Identification of the saponin extracted from Securidaca welwitschii. *Bulletin de la Société Royale des Sciences de Liège*, *40*, 498–501.

Effmert, U., Saschenbrecker, S., Ross, J., Negre, F., Fraser, C. M., Noel, J. P., Dudareva, N., & Piechulla, B. (2005). Floral benzenoid carboxyl methyltransferases: From in vitro to in planta function. *Phytochemistry*, *66*(11), 1211–1230.

Hamill, F. A., Apio, S., Mubiru, N. K., Bukenya-Ziraba, R., Mosango, M., Maganyi, O. W., & Soejarto, D. D. (2003). Traditional herbal drugs of Southern Uganda, II: Literature analysis and antimicrobial assays. *Journal of Ethnopharmacology*, *84*(1), 57–78.

Hedimbi, M., & Chinsembu, K. C. (2012). Ethnomedicinal study of plants used to manage HIV/AIDS-related disease conditions in the Ohangwena region, Namibia. *International Journal of Medicinal Plants Research*, *1*(1), 4–11.

Ji, J., Wang, Q., Wang, M., Chen, J., & Li, X. (2019). Chemical constituents from the stems of Securidaca inappendiculata Hassk. *Fitoterapia*, *137*, 104271.

Kadiri, A. B., Agboola, O. M., & Fashina, F. O. (2013). Ethnobotanical survey and phyto-anatomical studies of some common plants used for the treatment of epilepsy in some rural areas of South west Nigeria. *Journal of Pharmacognosy and Phytochemistry*, *2*(4), 175–182.

Katende, A., Ssegawa, P., & Tengnäs, B. (1995). *Useful trees and shrubs of Uganda. Technical Handbook series 10.* Regional Soil conservation Unit/SIDA.

Mabogo, D. E. N. (2012). *The ethnobotany of the Vhavenda*. University of Pretoria.

Maroyi, A. (2013). Traditional use of medicinal plants in south-central Zimbabwe: review and perspectives. *Journal of Ethnobiology and Ethnomedicine*, *9*(1), 31. https://doi.org/10.1186/1746-4269-9-31

Moeng, T. E. (2010). *An investigation into the trade of medicinal plants by muthi shops and street vendors in the Limpopo province.* MSc Dissertaion, University of Limpopo, South Africa.

Mongalo, N. I., McGaw, L. J., Finnie, J. F., & Staden, J. Van. (2015). Securidaca longipedunculata Fresen (Polygalaceae): A review of its ethnomedicinal uses, phytochemistry, pharmacological properties and toxicology. *Journal of Ethnopharmacology*, *165*, 215–226. https://doi.org/10.1016/j.jep.2015.02.041

Motlhanka, D. M. T., & Nthoiwa, G. P. (2012). Ethnobotanical survey of medicinal plants of Tswapong north, in eastern Botswana: a case of plants from Mosweu and Seolwane villages. *European Journal of Medicinal Plants*, *3*(1), 10–24.

Mustapha, A. A. (2013a). Ethno-medicinal field study of anti-fertility medicinal plants used by the local people in Keffi local government, Nasarawa State, Nigeria. *International Journal of Medicinal Plants Research*, *2*(5), 215–218.

Mustapha, A. A. (2013b). Ethno-medico-botanical uses of Securidaca longepedunculata (Fresen)(Family Polygalaceae) from Keffi local government, Nasarawa State, Nigeria. *Journal of Natural Remedies*, *13*, 133–137.

Nguta, J. M., Mbaria, J. M., Gakuya, D. W., Gathumbi, P. K., & Kiama, S. G. (2010). Antimalarial herbal remedies of Msambweni, Kenya. *Journal of Ethnopharmacology*, *128*(2), 424–432. https://doi.org/10.1016/j.jep.2010.01.033

Nordeng, H., Al-Zayadi, W., Diallo, D., Ballo, N., & Paulsen, B. S. (2013). Traditional medicine practitioners' knowledge and views on treatment of pregnant women in three regions of Mali. *Journal of Ethnobiology and Ethnomedicine*, *9*, 1–10.

Ochora, D. O., Kakudidi, E. K., Namukobe, J., Ipulet, P., Wakoli, D. M., Okore, W., Mwakio, E. W., Yeda, R. A., Cheruiyot, A. C., & Juma, D. W. (2021). Synergism in Antiplasmodial Activities of Artemether and Lumefantrine in Combination with Securidaca longipedunculata Fresen (Polygalaceae). *Plants*, *11*(1), 47.

Olajide, O. A., Oladiran, O. O., Awe, S. O., & Makinde, J. M. (1998). Pharmacological evaluation of Hoslundia opposita extract in rodents. *Phytotherapy Research: An International Journal Devoted to Pharmacological and Toxicological Evaluation of Natural Product Derivatives*, *12*(5), 364–366.

Onah C, C., Uzoma, A. I., Ottah, A, A.,Nnaedozie, C. L, Odeh J, O., & Omale, S. (2016). In vitro screening of the methanolic root extract of Securidaca welwitschii Oliv for anthelmintic activity. *World Journal of Pharmaceutical Research*, *8*(5), 263–269.

Oni, P. I., Jimoh, S. O., & Adebisi, L. A. (2014). Management of indigenous medicinal plants in Nigeria using phenological information. *Journal of Medicinal Plant Research*, *8*(16), 619–631.

Orwa, C., Mutua, A., Kindt, R., Jamnadass, R., & Simons, A. (2009). *Agroforestree Database: A tree reference and selection guide. Version 4. 2009.* World Agroforestry Centre. www.cifor-icraf.org/knowledge/publication/25648/?utm_source=cifor-icraf.org&utm_medium=site-search&utm_campaign=traffic-source

Plants of the World Online (2025). https://powo.science.kew.org

Semenya, S. S., & Potgieter, M. J. (2013). Ethnobotanical survey of medicinal plants used by Bapedi traditional healers to treat erectile dysfunction in the Limpopo Province, South Africa. *Journal of Medicinal Plants Research*, *7*(7), 349–357.

Sobiecki, J. F. (2008). A review of plants used in divination in southern Africa and their psychoactive effects. *Southern African Humanities*, *20*(2), 333–351.

Timité, G., Mitaine-Offer, A., Miyamoto, T., Tanaka, C., Paululat, T., Delaude, C., & Lacaille-Dubois, M. (2010). Presenegenin glycosides from Securidaca welwitschii. *Helvetica Chimica Acta*, *93*(11), 2237–2244.

van Wyk, B. E., & Gericke, N. (2000). *People's plants: A guide to useful plants of Southern Africa.* Pretoria: Briza Publications.

Van Wyk, B. E., Van Oudtshoorn, B., & Gericke, N. (2005). *Medicinal Plants of South Africa. 4th Impression.* Briza Publications, 254.

Victório, C. P., Carriço, J. B., & Lage, C. L. S. (2011). Polygala paniculata: A source of methyl salicylate produced through plant tissue culture. *Revista Ceres*, *58*(3), 269–272.

Viol, D. I. (2009). *Screening of Traditional Medicinal Plants from Zimbabwe for Photochemistry, Antioxidant, Antimicrobial, Antiviral and Toxicological Activities*. University of Zimbabwe.

Wang, D.-D., Li, Y., Wu, Y.-J., Wu, Y.-L., Han, J., Olatunji, O. J., Wang, L., & Zuo, J. (2021). Xanthones from Securidaca inappendiculata antagonized the antirheumatic effects of methotrexate in vivo by promoting its secretion into urine. *Expert Opinion on Drug Metabolism & Toxicology*, *17*(2), 241–250.

Chapter 5

The ecology, propagation, and conservation of *Securidaca longepedunculata* Fresen in Tropical East Africa

Collins E Bulafu and Mary Namaganda*
Department of Plant Sciences, Microbiology and Biotechnology
Makerere University, Kampala, Uganda

5.1 INTRODUCTION

5.1.1 Ecology

Securidaca longepedunculata, commonly known as the violet tree, is a small tree or shrub belonging to the family Polygalaceae. It has a wide distribution in tropical and subtropical Africa. In West Africa, it is native to Ghana, Senegal, Niger, Nigeria, Mali, Sierra Leone, Cameroon, and Cote d'Ivoire; in Central Africa it occurs in Cameron, Chad, Sudan, and Democratic Republic of Congo; in Southern Africa it occurs in Malawi, Mozambique, Namibia, South Africa, Zambia, and Zimbabwe (Orwa et al. 2009). In East Africa, its distribution ranges from wooded and bushed grasslands, scattered tree grasslands, open forest or woodland, and secondary bushland (Beentje and Ghazanfar 2007). The species' altitudinal expanse is approximately between 0 and 1,700 m above sea level (Orwa et al. 2009). The species is associated with sandy, acidic, and rocky soils (Tshisikhawe et al. 2012; Orwa et al. 2009). *S. longepedunculata* is widespread in Tropical East Africa (TEA). In Uganda and Tanzania, it is recorded from all floral regions of the two countries, i.e., four in Uganda and eight in Tanzania. In Kenya, it is less widespread, occurring mainly in central region of Mukweni and southern coastal area of Kwale. The species has multiple uses and is one of the most traded medicinal plants on the African continent (Tshisikhawe et al. 2012; Abubakar et al. 2019) and third most frequently sold herbal medicine species in St Balikuddembe market in Uganda (Tabuti et al. 2012). In west and southern Africa, extracts from the entire plant have been reported to treat a variety of ailments including malaria, urethral discharges, stomach problems, dysentery, rheumatism, and fibrositis (Tshisikhawe et al. 2012). In addition to medicinal uses, *S. longepedunculata* is used for firewood,

***Correspondence**: collinsabc@gmail.com

DOI: 10.1201/9781003544463-5

charcoal, poles, bee forage, ornamental, bows, fibre (inner bark), oil (flowers and seed), soap (roots), brooms (saplings), and fodder (Katende et al. 1995).

S. longepedunculata is facing multiple threats including unsustainable harvesting of its roots, bark and leaves for medicinal use, increasing habitat loss, fire, and climate change (Oni et al. 2014). Currently, little is known about the species distribution and abundance, limiting our understanding of ecological processing such as regeneration potential, and population size, thus making conservation planning a big challenge. Therefore, prediction and mapping of potential suitable habitats is important if we are to increase the species population and improve its conservation status. The availability and use of herbarium collections together with spatially explicit environmental data is fast becoming an important tool to address this critical gap through use of statistical models (Graham et al. 2004).

A variety of species distribution modelling tools that relate species occurrence, biophysical and environmental variables now exist and are widely used in many ecological applications, such as conservation and reserve planning, ecology, evolution, and invasive species management (Elith et al. 2006; Pearson et al. 2006; Welk et al. 2002). In Africa and TEA in particular, there is limited quality presence and absence data for species of limited distribution. In the face of these limitations, the choice of modelling techniques is skewed to only those that can utilize the available data (Graham et al. 2004). Even so, the usefulness of predictions from such models will depend on sample size (Wisz et al. 2008; Hernandez et al. 2006). Here we use georeferenced occurrence data from herbaria and environmental data from the WorldClim database to decipher the species' current and potential distribution envelope using a niche-based model Maximum entropy (MaxEnt) (Phillips et al. 2004).

5.1.2 Species description

S. longepedunculata is a small tree or shrub up to 6 m high, with a pale grey, smooth bark and oblong, more or less hairless alternate leaves that are variable in size and shape and crowded towards the stem tips. The flowers are sweet scented, small, about 1 cm in length, pink to lilac or purple in colour, and are clustered (van Wyk and Albrecht 2008). The flowers are produced in the months of January, March, April, September, October, November, and December. Fruits are dry and indehiscent, heavily veined, occasionally smooth, oblong, purplish green when young and possess a membranous wing of about 4 cm long. They appear in the months of May, June, and September.

The fruit is a samara with a round seed of about 8–10 mm wide (Figure 5.1). The seed is recalcitrant and is enclosed within a hard pericarp. Some researchers have reported the seed to have a hard seed coat and thus termed the seed dormant, reporting the need to de-coat the seed to improve

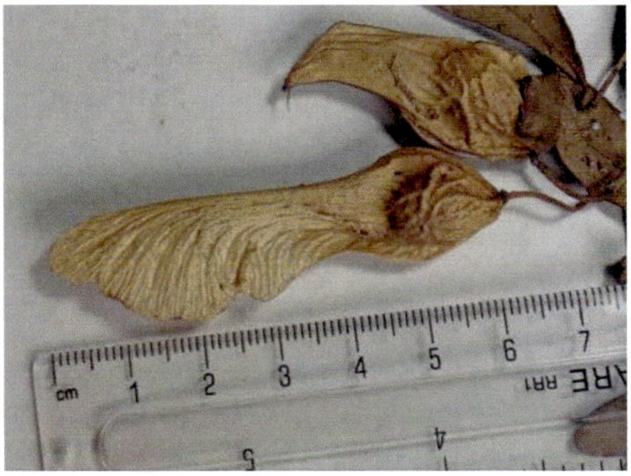

Figure 5.1 Dry fruit of *S. longepedunculata.*

germination rates. The 'de-coating' referred to is the removal of the pericarp (Tiawoun et al. 2017). It is important to understand the structure of the fruit of *S. longepedunculata* because being samaras, the fruits are dry and indehiscent; thus, the fruit is sometimes wrongly referred to as the seed. In nature, effective wear and tear of the hard fruit wall is vital before the seed can be exposed for germination, thus the low germination potential that is reported. Handbook series of the 'Useful Trees and Shrubs' reports that the seeds remain viable for a long period of time at room temperature if kept dry (Mbuya et al. 1994; Katende et al. 1995; Bein et al. 1996; Bekele-Tesemma 2007). However, it is possible that these publications refer to the fruit (samara) because the seed of *S. longepedunculata* is recalcitrant and thus sensitive to desiccation (Kaniaru et al. 2021). Indeed, it would be important to investigate the precise longevity period of the fruits and seeds while in storage. The presence of moisture in nature would be important in supporting the rotting of the hard pericarp and at the same time allow for seed germination but still, low germination potential is reported. The recalcitrant seeds of *S. longepedunculata* possibly have very low longevity and perhaps the time it takes for the fruit wall to rot is too long for the seed to maintain its viability. In Uganda and Tanzania, the seeds are light, reported weight being 36,000–36,500 seeds/kg compared to 6,000 seeds/kg in Ethiopia and Eritrea (Mbuya et al. 1994; Katende et al. 1995; Bein et al. 1996; Bekele-Tesemma 2007).

5.1.3 Species distribution

The species has a wide distribution in tropical Africa. In Eastern Africa, it has been recorded in Kenya, Tanzania, Ethiopia, Burundi, Rwanda, and Uganda (Baloyi and Tshisikhawe 2009; Tshisikhawe et al. 2012). In Uganda and Tanzania, it is known from the four and eight floral regions, respectively, while in Kenya, it has limited distribution to the far south-eastern coastal areas. The species is mainly encountered in lowland dry woodland and savanna regions and occasionally in forests. It grows within an altitude range of 0–1,700 m, with a mean annual rainfall of 600–1000 mm (Orwa et al. 2009). In Uganda, the species is locally known as Elilyoi (Ateso), Lilo (Luganda), Oiyofe (Lugbara), Loloyi (Lugwere), Wadambasima (Lugisu), Mukondwa (Lusoga), Mweya (Runyankole), and Nkondwe (Runyoro) (Katende et al. 1995). However, in other African countries, it also has various names, for example, Umfufu (Ndebele), Jodo (Mandinka), es a manahi (Amharic), Ipeta (Yoruba), Muteya, Mzigi, Chipvufana, and Mufufu (Swahili) (Mongalo et al. 2015). In Tanzania, the species is called Furudau (Fiome), Mteyo, and Mteyu (Nyam).

5.1.4 Ethnobotanical use

The species has a variety of uses, such as medicine, timber, and firewood among others, and continues to be targeted by people for these values in the areas where it is native, thus threating its long-term survival. For instance, the violet tree, *S. longepedunculata*, is known as the king of traditional medicinal plants in South Africa, where it is used to treat almost every conceivable ailment. The species is heavily harvested from the wild for various ethnobotanical uses. It has been documented to treat various ailments, for example, coughs, fever, malaria, hernias, gonorrhea, palpitations, headaches, oedema (Chhabra et al. 1991; Maroyi 2013), rheumatism (Mongalo et al. 2015), diabetes, and sexual impotence (Ogunmefun and Gbile 2012). The roots are targeted for preparation of the remedies, putting the species health at risk (Abubakar et al. 2019; Oluwaniyi et al. 2022). The species has been used in unique ways, such as declaring a person dead and protection from evil spirits (Bamogo et al. 2023). In Ethiopia and South Africa the species is also harvested as a food, particularly leaves, and source of other materials such as building poles, essence from the roots, soap from the bark, and fibre from the inner bark to make ropes and veterinary medicine (Bekele-Tesemma 2007; Orwa et al. 2009). In Tanzania the species is used for firewood, building poles, fodder, bee forage, ornamental and fibre from the inner bark, soap from the roots, and oil from flowers and seeds, as well as brooms (Mbuya et al. 1994; Katende et al. 1995). The species also has been recommended and used extensively in apiculture due to its scented flowers, particularly in Eritrea. The flowers have also been exploited for lipid production and used to stain furniture.

5.1.5 Aim

S. longepedunculata is an important medicinal species which is facing multiple threats from unsustainable harvesting and habitat loss which threatens its population. Further, *S. longepedunculata* regeneration in the wild is poor yet *ex situ* propagation by seed would be difficult without pretreatment. Therefore, there was a need to review the literature on ecology and propagation, and model its distribution to inform conservation strategies. The aim was to assess the ecology and management practices of *S. longepedunculata* in TEA. The objectives were to model the distribution of *S. longepedunculata*, review information on the species seed propagation, management, species threats, and conservation potential.

5.2 METHODS

5.2.1 Study area

TEA, defined as the regions of Uganda, Kenya, and Tanzania, is one of the most biodiversity hotspots in Africa, accounting for about a quarter of tropical plants in Africa (Moshi et al. 2010). This region occupies an area of 1,768,706 km^2 and extends from 5°2'N to 11°45'S and 29°20'E to 41°54'E (Figure 5.2). We chose these three countries for modelling as we could retrieve four or more occurrence records for *S. longepedunculata* from public databases for these countries and less than 4 for other countries bordering TEA.

5.2.2 Current distribution of *S. longepedunculata*

We used MaxEnt to model the current species distribution.

5.2.3 MaxEnt model description

MaxEnt 3.4.4 was used to model the current species distribution based on the species occurrence data harvested from herbarium collections and bioclimatic data (Phillips et al. 2006). MaxEnt is a general-purpose machine learning presence-only model based on a precise and straightforward mathematical formulation that uses predictor datasets that can be categorical or continuous in nature to distinguish species occurrence patterns (Akpoti et al. 2020; Merow at al. 2013; Heumann et al. 2013). MaxEnt was chosen because it has the best predictive power across all sample sizes including small sample sizes and uses presence data only compared to other methods such as GBM, Genetic Algorithm for Rule-Set Prediction, Domain, BRUTO (which fits a generalized additive model using an adaptive back-fitting procedure) (Brook et al. 2006; Wisz et al. 2008; Hernandez et al. 2006; Phillips et al. 2006). MaxEnt is also user-friendly with its graphical user interface and command-line functions and provides useful model assessment tools such as jackknife environmental parameter contributions, species-environment curves, and Area Under the Curve of the Receiver Operating Characteristic

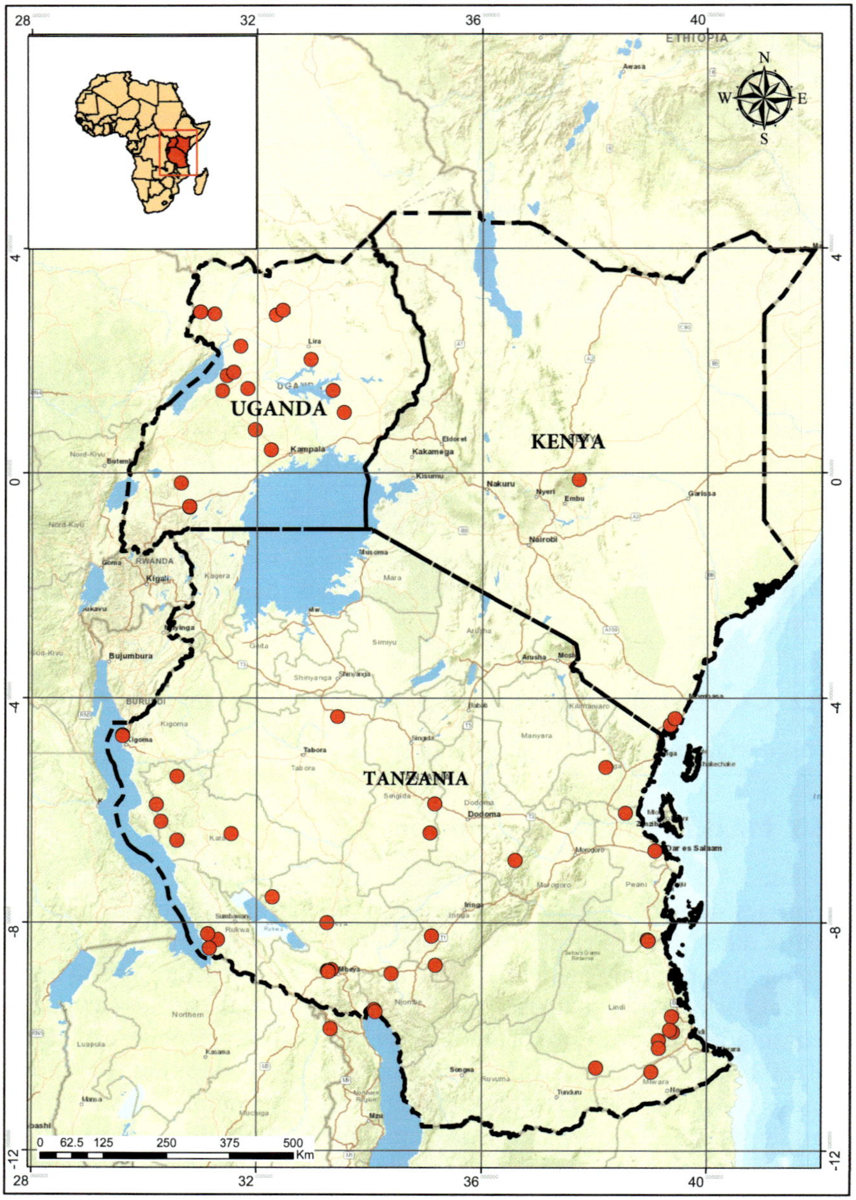

Figure 5.2 Spatial distribution *of S. longepedunculata* occurrence points in TEA.

as a metric of model performance (Phillips et al. 2006). MaxEnt has been used to infer species environmental requirements from conditions at known occurrences, making it a powerful tool in conservation planning, understanding biodiversity, and the processes that drive them.

5.2.4 Data sources

5.2.4.1 *Suitability of species for modelling*

S. longepedunculata holds two unique characteristics that make it a good candidate to model its distribution using MaxEnt. Firstly, it shows widespread geographic distributions with clear ecological/environmental patterns. Secondly, there are a reasonable number of georeferenced occurrence localities for the species based on confirmed herbarium specimens.

5.2.4.2 *Species occurrence data*

Species occurrence data were extracted from the Missouri Botanical Garden Herbarium dataset and Makerere University Herbarium (MHU) as of September 2024. The species '*Securidaca longepedunculata*' was searched for in the Missouri Botanical Garden, Tropicos Home page (www.tropicos.org/home) and extracted entries for Uganda, Kenya, and Tanzania under the specimens tab. We retrieved 50 occurrence records, that is, 30 records for Tanzania, 3 for Kenya, and 17 for Uganda, from the www.tropicos.org/home and MHU (Figure 5.2). The occurrence records were converted into CSV format files for MaxEnt modelling.

5.2.4.3 *Environmental variables*

We used bioclimatic variables for this analysis (Figure 5.2). These were downloaded from the WorldClim global database (www.worldclim.org) at a spatial resolution of 30 seconds. We used climatological data from 1970 to 2000. They describe the mean monthly values of various variables, which were processed and converted to ASCII raster grid format, as required by MaxEnt. Firstly, all 19 environmental variables were included in MaxEnt model building to obtain the variable importance by the jackknife test. Then the variable percentage contribution of ≤2% was used as the basis for the elimination of environmental variables in the final analysis (Wang et al. 2024; Shi et al. 2021). The percentage contribution and permutation importance for the nine variables that contributed most to the model fitting are reported.

5.2.4.4 *Model building*

For *S. longepedunculata*, because we did not have a large data set of occurrences, we chose zero random partitions of the occurrence localities.

Otherwise, we ran MaxEnt on default settings. The algorithms were run with all 19 bioclimatic variables. The selected species occurrence data and environmental factors were entered into MaxEnt 3.4.4 to model the current distribution (http://biodiversityinformatics.amnh.org/open_source/maxent). The following settings were used: Number of training samples = 50; random test percentage = 0; regularization multiplier = 1; the maximum number of background points = 10 000, replicates = 10; number of iterations = 500 (Phillips et al. 2006); the Cloglog transform as a continuous index of habitat suitability was selected as the output format. The area under the curve (AUC) of the reciever operating characteristic (ROC) was used to evaluate the fit of the MaxEnt model to the true presence and absence data (Heumann et al. 2013). If AUC ≤ 0.5, it indicates a random prediction, while AUC > 0.5 indicates a better model prediction. The output ASCII grid produced by MaxEnt is continuous probability data ranging from 0 to 1, which represents the habitat suitability of a species in a specified region. A 0.77 was adopted as the threshold to transform the probability data into binary data (0/1). Grids with probability higher than 0.771 were defined as suitable habitats (Song et al. 2019).

5.2.4.5 *Model evaluation*

The MaxEnt model outputs a map of occurrence probabilities and tables of model selection (e.g., variable contribution to the model) and the AUC for the training and validation datasets. Variable contributions and AUC were displayed as jackknife plots. The continuous probability maps were then converted into binary maps (suitable vs unsuitable) based on the probabilities being equal and that the model was correctly classified as (i) highly suitable, (ii) moderately suitable, (iii) low suitability, and (iv) unsuitable areas. To do this, each pixel on the potential distribution map for *S. longepediculata* was ranked from 0 to 1; these pixels were then grouped into four types of potential habitat: 'highly suitable' (> 0.941), 'moderately suitable' (0.86–0.940), 'low suitability' (0.78–0.859), and 'not suitable' (< 0.77) habitats.

5.2.5 Seed propagation, species threats, and conservation status

Literature on seed propagation, threats, and conservation of the species was reviewed.

5.3 RESULTS

5.3.1 Current species distribution

The jackknife plots from the MaxEnt model were used to determine the contribution of nine environmental variables to the final maps produced.

Table 5.1 WorldClim bioclimatic variables and their corresponding percent contribution and permutation importance

Climatic Variable	*Description*	*Percent Contribution*	*Permutation Importance*
Bio01	Annual Mean Temperature	≤2	0
Bio02	Mean Diurnal Range (Mean of monthly (max temp – min temp)	6.6	7.6
Bio03	Isothermality (BIO2/BIO7) (×100)	**7.2**	**14**
Bio04	Temperature Seasonality (standard deviation ×100)	7.4	5.7
Bio05	Max Temperature of Warmest Month	2	6.1
Bio06	Min Temperature of Coldest Month	≤2	1,6
Bio07	Temperature Annual Range (BIO5-BIO6)	3.4	3
Bio08	Mean Temperature of Wettest Quarter	≤2	0.1
Bio09	Mean Temperature of Driest Quarter	**≤2**	**4.4**
Bio10	Mean Temperature of Warmest Quarter	3.3	4.9
Bio11	Mean Temperature of Coldest Quarter	≤2	0
Bio12	Annual Precipitation	**53.5**	**43.9**
Bio13	Precipitation of Wettest Month	≤2	0.1
Bio14	Precipitation of Driest Month	4.8	5.5
Bio15	Precipitation Seasonality (Coefficient of Variation)	≤2	2.3
Bio16	Precipitation of Wettest Quarter	≤2	0.1
Bio17	Precipitation of Driest Quarter	≤2	0
Bio18	Precipitation of Warmest Quarter	≤2	1.8
Bio19	Precipitation of Coldest Quarter	11.8	9.3

Note: The dominant environmental factors are shown in bold.

The distribution of *S. longepedunculata* is driven by different environmental factors in TEA. The most important environmental variable influencing *S. longepedunculata* distribution was predominantly rainfall and to a lesser extent temperature-related. Annual precipitation (Bio12), precipitation of the coldest quarter (Bio19), and isothermality (Bio3) were the dominant factors driving the species distribution regardless of whether training gain, test gain, or AUC is used (Figure 5.3). Under current conditions, the average training AUC was 0.8803, and the average test AUC was 0.8166 ± 0.0679.

The most suitable areas for natural populations of *S. longepedunculata* in Uganda are West Nile, Acholi, Teso, Elgon, and south-western sub-regions; in Kenya it is found in the central region around Mukweni and Kwale, and a thin belt around far southern coastline; and the following regions in Tanzania: Rukwa, Mbeya, Iringa, Ruvuma, the coastal regions of Pwani, Lindi, and Mtwara (Figure 5.4).

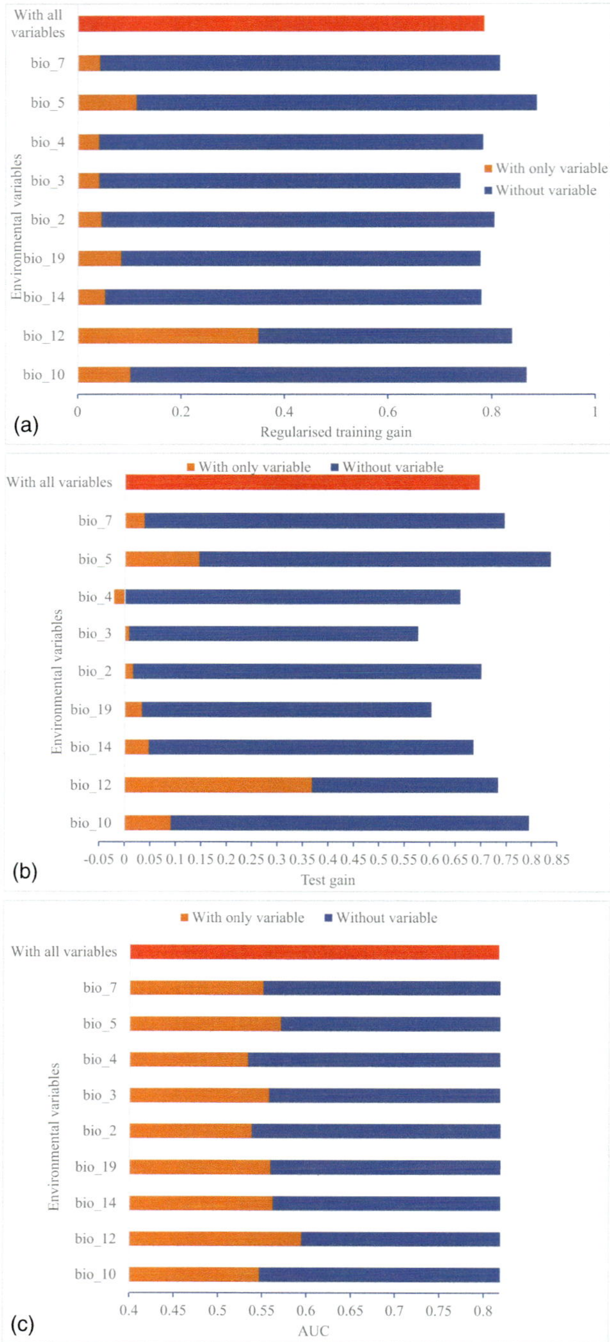

Figure 5.3 Jackknife test of the importance of environmental variables for TEA. (a) Regularized training gain, (b) test gain, and (c) AUC value.

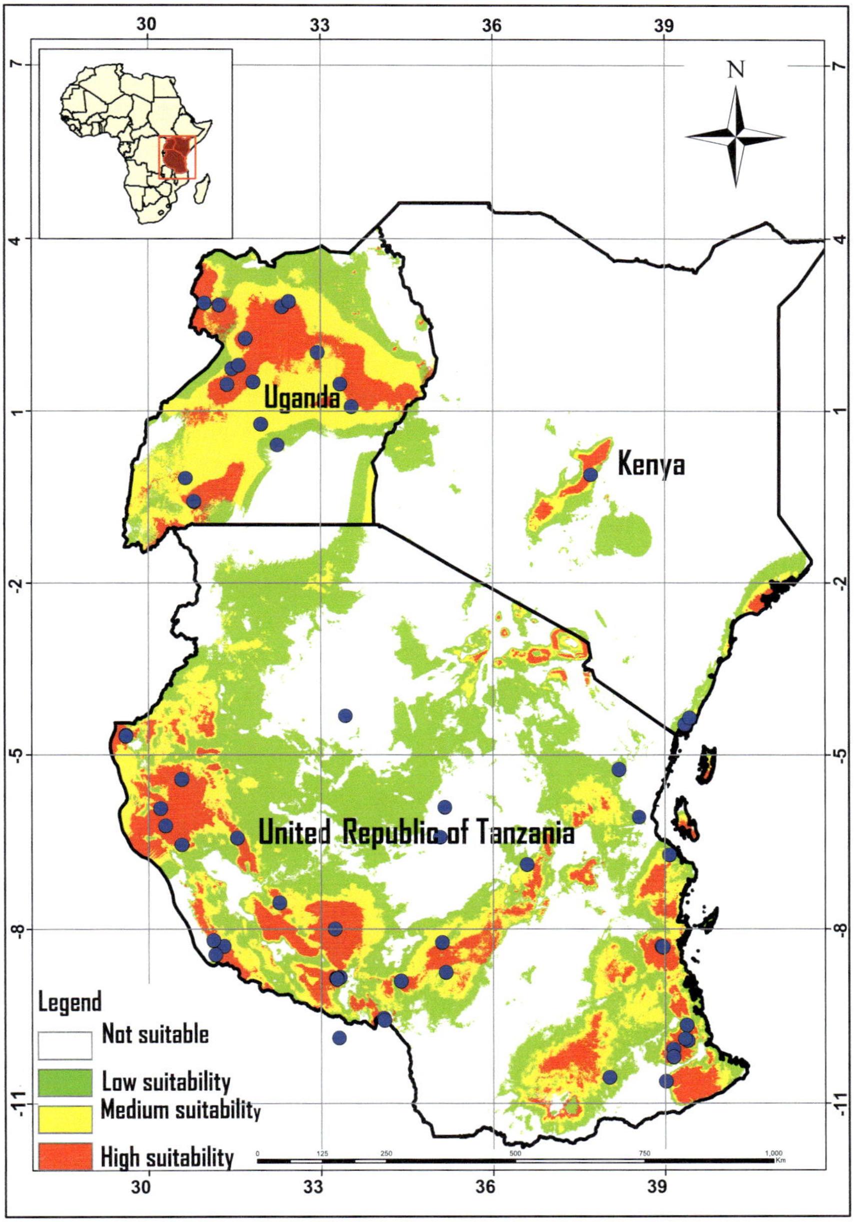

Figure 5.4 Potentially suitable areas for *S. longepedunculata* under current climate conditions in TEA. The blue cycle represents the occurrence records of the species.

5.4 DISCUSSION

5.4.1 Species distribution

The most important environmental variables influencing the distribution of the current potentially suitable areas for *S. longepedunculata* in TEA are predominantly rainfall and, to a less extent, temperature. For instance, annual precipitation (Bio12), precipitation of the coldest quarter (Bio19), and isothermality (Bio3) were the dominant factors driving the species distribution. The factors influencing the current suitable areas for *S. longepedunculata* in TEA are interesting and at odds with other studies. Studies on other genera such as *Solanum* have shown temperature and rainfall to be among the most important factors influencing their distribution (Kawuma et al. 2025; Shi et al. 2021). Studies on *Magnolia wufengensis,* a critically endangered species with a limited distribution range and sparse population, report that low temperature variables (e.g., Bio11 and Bio6) were the most important environmental variables affecting its distribution globally (Shi et al. 2021). In another study, temperature was found to be a major determinant for the distribution of *M. officinalis.* ssp. *officinalis* yet for *M. officinalis.* ssp. *biloba*, precipitation rather than temperature was more important (Song et al. 2019). More generally, rainfall and temperature are the major factors that influence the distribution of vegetation; in particular, maximum and minimum rainfall and average temperature greatly limit the distribution and dispersal of vegetation (Thuiller et al. 2003; Crawford 2008).

5.4.2 Current suitability mapping

The high AUC for TEA means that the model had good prediction and explanatory power. The most suitable areas for natural populations of *S. longepedunculata* in Uganda are West Nile, Acholi, Teso, Elgon, and south-western sub-regions; in Kenya it is found in central region around Mukweni and Kwale, and a thin belt around far southern coast line; and the following regions in Tanzania: Rukwa, Mbeya, Iringa, Ruvuma, the coastal regions of Pwani, Lindi, and Mtwara. The fact that majority of highly suitable areas fall outside protected areas implies that the species will continue to come under pressure. At the same time, a small fraction of suitable areas falls within the protected areas, i.e., Murchison Falls National Park which presents us with an opportunity for species conservation and perhaps restoration. The areas where the species previously occurred and are now modelled as unsuitable are concerning. This might mean three things: Firstly, these areas have been converted into other land uses and therefore the environmental conditions are unsuitable for this *S. longepedunculata*. Secondly, there is a shift in species range particularly in Uganda and parts of Tanzania, with some of the areas ranked as unsuitable by the model based on current conditions even when there were occurrences there from historical data. Thirdly, our dataset is too small to cover the entire environmental envelope for this species. This might be particularly true for Kenya where we only managed to map species using only three occurrence records.

5.4.3 Seed propagation and management practices

5.4.3.1 Seed germination

The seeds need to be sown immediately after collection and after removal of the stony pericarp. Handling protocols for recalcitrant seeds need to be followed to ensure that the seeds do not dry and remain viable throughout the preparation period. It is important to ensure the collection of mature seeds and these can be detected by colour change of the fruit, from green with a purple tinge to straw colour, and a percent moisture content of 35±1 (Kaniaru et al. 2021). Under natural conditions, the seeds have very low germination rates (Tiawoun et al. 2017). Under controlled conditions, most researchers report the need for some form of pretreatment in order to break seed dormancy, including removal of the seed coat, soaking in cold water for 24 hours before planting, immersion in hot water, soaking in sulfuric acid, and using growth regulating hormones (Zulu et al. 2011). Removal of the seed coat (actually pericarp) is recommended as it was found to result in the highest germination percentage (Zulu et al. 2011; Mulumbati et al. 2023; Tiawoun et al. 2017). It is possible that non-physical seed dormancy-breaking agents, like acids and hot water, are likely to kill the embryo, yet maintaining the seeds moist after removal of the pericarp would be sufficient to keep the seeds alive. This could explain why experiments on comparative seed dormancy-breaking protocols have recommended the removal of the seed coat as the most effective pretreatment technique. There is a long period reported for when the seed remains viable; however, the precise longevity period of the fruits and seeds while in storage is not well understood.

5.4.3.2 Seed management practices

It is recommended to sow clean seeds; rinse the seeds in clean fresh water and soak for 2–4 hours. Seedlings sprout between 8 and 20 days but can take longer, and germination occurs after 4–5 weeks. Germination success depends on the type of medium used for sowing. These could be vermiculite, fine or coarse grit, sawdust, coco peat, forest soil, fine sand, or habitat soil. The medium should be well drained and maintained at 30°C and seedlings transplanted when at the two-leaf stage (Anjarwalla et al. 2016; Kaniaru et al. 2021).

In vitro protocols have been proposed (Lijalem and Feyissa 2020; Mulumbati et al. 2023). These have their disadvantages, particularly to poor farmers in TEA, including the high cost of implementation and the requirement of skilled personnel. The protocols involve the selection of only those genotypes with desirable traits and this tends to lower the genetic diversity within the local populations. On the other hand, germination success using conventional methods can be improved with the use of germination incubators or the construction of non-mist propagators that are of low cost (Kaniaru et al. 2021).

5.4.3.3 Seed survival

In nature, seedlings of *S. longepedunculata* are reported to have a high mortality rate. When *ex situ* propagation is successful, transplanting is not recommended because the long tap roots of the seedlings break easily (Mbuya et al. 1994; Katende et al. 1995; Bein et al. 1996; Bekele-Tesemma 2007) but can be done in the early stages of development when the tap root is still short. It is recommended to either plant in deep pots or sow directly *in situ*. The seeds of *S. longepedunculata* grow better at a soil depth of 4 cm between 20 °C and 30 °C (Baloyi and Tshisikhawe 2009).

5.4.4 Current threats

Firstly, the greatest threat to the species comes from its medicinal use. The leaves, bark, seeds, and roots are harvested unsustainably for medicinal use, making its regeneration difficult (Abubakar et al. 2019; Maroyi 2013). Secondly, the species is targeted for charcoal and firewood production. Thirdly, habitat conversion to agricultural systems and urbanization are currently shrinking the species' distributional range. For instance, in Uganda, the species is now only predicted to occur in small areas in West Nile and almost absent in the previously favourable habitats in Central Uganda.

5.4.5 Conservation

The species faces multiple threats including over-exploitation particularly the unsustainable root harvesting for medicinal and non-medicinal uses. Further its natural habitat continues to shrink from the finite pressure on land, such as urbanization and agricultural expansion (Abubakar et al. 2019). In other areas, fire is used for the management of savanna ecosystems, yet *S. longepedunculata* may not survive as it lacks fire-tolerant traits (such as smooth barks, generally short tree (2–6 m) with dropping branches). The species enjoys protection in many African countries, such as South Africa, where it has been elevated to protected status at provincial and national levels (Government of South Africa 1998). In some countries outside TEA, the species enjoys protection. In South Africa, for example, the species is nationally protected under the National Forest Act of 1998 and deliberate efforts to include the species among the agroecosystems are one way to enhance its protection. While the species enjoys protection within gazetted areas, its survival outside these areas may depend on the efforts of the communities. In many other African countries including Uganda, there is no special conservation effort to protect this over-exploited species. Thus, given its wealth of beautiful blooms, *S. longepedunculata* could be promoted as an ornamental tree.

Although *S. longepedunculata* is reported to be a fairly fast-growing species, it remains threatened in nature because it has a low regeneration potential that is caused by poor seed germination rates and over-exploitation. It prefers light, well-drained weakly acidic soils (pH 5–7), tolerates low fertility and can survive bush fires (https://tropical.theferns.info/viewtropical.php?id=Securidaca+longipedunculata). In Kenya, flowers are produced during March–April and fruits appear in May and hang on the trees for many months (flower and seed between April and August). *S. longepedunculata* is propagated by seed but can also grow from root suckers. The tree's beautiful pink and scented flowers make it a good candidate as an ornamental plant in urban areas landscaping as well as in gardens. Further, the tree's scented flowers are a bee haven; together this can be good ideas for conservation solutions. For instance, in Eritrea, the tree is one of the most valuable lowland honey sources. Therefore, in areas where apiculture is an economic activity, strategic tree planting drives to increase honey production are a worthy intervention to safeguard the species. The species enjoys protection in some African countries such as South Africa (Moeng, 2010).

5.5 CONCLUSION

The most suitable areas for natural populations of *S. longepedunculata* in Uganda are West Nile, Acholi, Teso, Elgon, and south-western subregions; in Kenya it is found in the central region around Mukweni and Kwale, and a thin belt around far southern coastline; and the following regions in Tanzania: Rukwa, Mbeya, Iringa, Ruvuma, the coastal regions of Pwani, Lindi, and Mtwara. This distribution is influenced by environmental variables, mainly precipitation and, to a lesser extent, temperature fluctuation. The majority of suitable areas lie outside protected areas; to a lesser extent, some areas fall within protected areas: Murchison Falls and Mt. Elgon National Parks in Uganda and Rukwa, Lukwati, and Selous Game Reserves in Tanzania. Given that most of the suitable habitats lie outside formal protected areas, deliberate efforts need to be made to produce seedlings and plant both in the wild and on farms. Propagators need to understand that low germination rates of this species are because of the recalcitrant nature of the seeds that are protected in a tough fruit wall but not seed dormancy due to a hard seed coat as widely believed.

REFERENCES

Abubakar, Umar, Habib Danmalam, Hajara Ibrahim, and Bb Maiha. 2019. "A Review on African Violet Tree (Securidaca Longipedunculata): A Traditional Drug with Multiple Medicinal Uses." *Special Journal of Chemistry and Biochemistry Innovations* 4 (October): 7–14.

Akpoti, Komlavi, Amos T. Kabo-bah, Elliott R. Dossou-Yovo, Thomas A. Groen, and Sander J. Zwart. 2020. "Mapping Suitability for Rice Production in Inland Valley Landscapes in Benin and Togo Using Environmental Niche Modeling." *Science of the Total Environment* 709 (March). https://doi.org/10.1016/J.SCITOTENV.2019.136165.

Anjarwalla, Parveen, Steven R. Belmain, Phosiso Sola, R. Jamnadass, and Philip C. Stevenson. 2016. *Handbook on Pesticidal Plants*. Nairobi: World Agroforestry Centre (ICRAF).

Baloyi, O., and M. P. Tshisikhawe. 2009. "The Population Ecology of Securidaca longepedunculata Fresen in Nylsvley Nature Reserve, Limpopo Province, RSA." *South African Journal of Botany* 75 (2): 430. https://doi.org/10.1016/j.sajb.2009.02.134.

Bamogo, Rabila, Achille Sindimbasba Nikièma, Mamounata Belem, Massamba Thiam, Youssouph Diatta, and Roch Kounbobr Dabiré. 2023. "Cross-Sectional Ethnobotanical Survey of Plants Used by Traditional Health Practitioners for Snakebite Case Management in Two Regions of Burkina Faso." *Phytomedicine Plus* 3 (3). https://doi.org/10.1016/j.phyplu.2023.100471.

Beentje, H., and S. A. Ghazanfar. 2007. *Flora of Tropical East Africa (Polygalaceae)*. England: Royal Botanic Gardens, Kew.

Bein, E., Habte, B., Jaber, A., Birnie., A and Tengnas, B. 1996. *Useful Trees and Shrubs in Eritrea. Identification, Propagation and Management for Agricultural and Pastoral Communities*. Nairobi: Regional Soil Conservation Unit, RSCU, 1996. (Regional Soil Conservation Unit, RSCU, Technical Handbook Series; 12).

Bekele-Tesemma, A. 2007. *Useful Trees and Shrubs for Ethiopia: Identification, Propagation and Management for 17 Agroclimatic Zones*. Edited by S. Maundu, P. Tengnas, B. Kelbessa, E. Demissew. Nairobi: RELMA in ICRAF Project.

Brook, Barry W., Corey J. A. Bradshaw, Lian Pin Koh, and Navjot S. Sodhi. 2006. "Momentum Drives the Crash: Mass Extinction in the Tropics." *Biotropica* 38 (3): 302–5. https://doi.org/10.1111/j.1744-7429.2006.00141.x.

Chhabra, S. C., R. L. A. Mahunnah, and E. N. Mshiu. 1991. "Plants Used in Traditional Medicine in Eastern Tanzania. V. Angiosperms (Passifloraceae to Sapindaceae)." *Journal of Ethnopharmacology* 33 (1): 143–57. https://doi.org/10.1016/0378-8741(91)90173-B.

Crawford, R. M. M. 2008. "Plants at the Margin: Ecological Limits and Climate Change." *Plants at the Margin: Ecological Limits and Climate Change*, January, 1–478. https://doi.org/10.1017/CBO9780511754906.

Elith, Jane, H. Graham, Catherine, Robert P. Anderson, Miroslav Dudík, Simon Ferrier, Antoine Guisan, et al. 2006. "Novel Methods Improve Prediction of Species' Distributions from Occurrence Data." *Ecography* 29 (2): 129–51. https://doi.org/10.1111/j.2006.0906-7590.04596.x.

Government of South Africa. 1998. "National Forests Act, 1998." *Government Gazette* 400 (19408): 2–82. www.gov.za/sites/default/files/gcis_document/201409/a84-98.pdf.

Graham, Catherine H., Simon Ferrier, Falk Huettman, Craig Moritz, and A. Townsend Peterson. 2004. "New Developments in Museum-Based Informatics and Applications in Biodiversity Analysis." *Trends in Ecology and Evolution* 19 (9): 497–503. https://doi.org/10.1016/J.TREE.2004.07.006.

Hernandez, Pilar A., Catherine H. Graham, Lawrence L. Master, and Deborah L. Albert. 2006. "The Effect of Sample Size and Species Characteristics on Performance of Different Species Distribution Modeling Methods." *Ecography* 29 (5): 773–85. https://doi.org/10.1111/j.0906-7590.2006.04700.x.

Heumann, Benjamin W., Stephen J. Walsh, Ashton M. Verdery, Phillip M. McDaniel, and Ronald R. Rindfuss. 2013. "Land Suitability Modeling Using a Geographic Socio-Environmental Niche-Based Approach: A Case Study from Northeastern Thailand." *Annals of the Association of American Geographers* 103 (4): 764–84. https://doi.org/10.1080/00045608.2012.702479.

Kaniaru, Gerald, Collins Wafula, Samuel Kiboi, Jonathan Sila. 2021. Medicinal Plant Species of Makueni, Kenya: Securidaca longepedunculata Fresen A Guide to Seed Collection, Processing, Germination, Propagation and Nursery Practices. *National Museums of Kenya*, Volume 1, pp. 1–5.

Katende, A.B, A. Birnie, and B. Tengnas. 1995. *Useful Trees and Shrubs for Uganda Identification, Propagation and Management for Agricultural and Pastoral Communities*. Edited by B. Tengnas. Nairobi: Regional Land Management Unit, RELMA/Sida, ICRAF House, Gigiri. https://apps.worldagroforestry.org/usefultrees/frontpages/Useful_Trees_Uganda.pdf.

Kawuma, Carol, Brita Stedje, Esther Katuura, Perpetua Ipulet, John Wasswa Mulumba, Esezah Kakudidi, Felly Tusiime Mugizi, Anifat Olayemi Bello, and Mary Namaganda. 2025. "Diversity, Distribution and Habitat Suitability Assessment of Solanum Species of the Leptostemonum Clade in Uganda." *African Journal of Ecology* 63 (2): e70021. https://doi.org/10.1111/aje.70021.

Lijalem, Teklebrihan, and Tileye Feyissa. 2020. "In Vitro Propagation of Securidaca longipedunculata (Fresen) from Shoot Tip: An Endangered Medicinal Plant." *Journal of Genetic Engineering and Biotechnology* 18 (December): 3. https://doi.org/10.1186/s43141-019-0017-0.

Maroyi, Alfred. 2013. "Traditional Use of Medicinal Plants in South-Central Zimbabwe: Review and Perspectives." *Journal of Ethnobiology and Ethnomedicine* 9 (May): 31. https://doi.org/10.1186/1746-4269-9-31.

Mbuya, L. P., H. P. Msanga, C. K. Birnie, A. Birnie, and B. Tengnas. 1994. *Useful Trees and Shrubs for Tanzania: Identification, Propagation and Management for Agricultural and Pastoral Communities*. Edited by Paul Rimmerfors/RSCU. Sweden: Regional Soil Conservation Unit (RSCU) Swedish International Development Authority (SIDA) Embassy of Sweden. https://apps.worldagroforestry.org/usefultrees/frontpages/Useful_Trees_Tanzania.pdf.

Merow, Cory, Matthew J. Smith, and John A. Silander. 2013. "A Practical Guide to MaxEnt for Modeling Species' Distributions: What It Does, and Why Inputs and Settings Matter." *Ecography* 36 (10): 1058–69. https://doi.org/10.1111/J.1600-0587.2013.07872.X.

Moeng, T.E. 2010. An investigation into the trade of medicinal plants by muthi shops and street vendors in the Limpopo Province, South Africa. Master of Science Dissertation, University of Limpopo, South Africa.

Mongalo, N. I., L. J. McGaw, J. F. Finnie, and J. Van Staden. 2015. "Securidaca longipedunculata Fresen (Polygalaceae): A Review of Its Ethnomedicinal Uses, Phytochemistry, Pharmacological Properties and Toxicology." *Journal of Ethnopharmacology* 165: 215–26. https://doi.org/10.1016/j.jep.2015.02.041.

Moshi, Mainen, Ester Innocent, Joseph Magadula, Donald Otieno, Anke Weisheit, Pamela Mbabazi, and Ramadhani Nondo. 2010. "Brine Shrimp Toxicity of Some Plants Used as Traditional Medicines in Kagera Region, North Western Tanzania." *Tanzania Journal of Health Research* 12 (July): 63–67. https://doi.org/10.4314/thrb.v12i1.56287.

Mulumbati, Magnifique, Mario Godoy, Louis Longanza, Jan Bogaert, Stefaan Werbrouck, Yannick Useni Sikuzani, and Mazinga Kwey Michel. 2023. "In Vitro Regeneration Protocol for Securidaca longepedunculata Fresen, a Threatened Medicinal Plant within the Region of Lubumbashi (Democratic Republic of the Congo)." *Conservation* 3 (August): 411–25. https://doi.org/10.3390/conservation3030028.

Mutke, Jens, G. Kier, G. Braun, Chr Schultz, and W. Barthlott. 2001. "Patterns of African Vascular Plant Diversity – A GIS-Based Analysis." *Systematics and Geography of Plants* 71 (2): 1125–36. https://doi.org/10.2307/3668744.

Ogunmefun, Olayinka, and Z Gbile. 2012. "An Ethnobotanical Study of Anti-Rheumatic Plants in South Western States of Nigeria." *Asian Journal of Science and Technology* 4 (November): 063–066.

Oluwaniyi, Omolara O., Ademidun A. Adesibikan, and Stephen S. Emmanuel. 2022. "Evaluation of Wound-Healing Activity of Securidaca longepedunculata Root Extract in Male Wistar Rats." *ChemistrySelect* 7 (26): e202200711. https://doi.org/10.1002/slct.202200711.

Oni, Peter, S. Jimoh, and L. A. Adebisi. 2014. "Management of Indigenous Medicinal Plants in Nigeria Using Phonological Information." *Journal of Medicinal Plant Research* 8 (April): 619–31. https://doi.org/10.5897/JMPR2013.5108.

Orwa, C., A. Mutua, R. Kondt, R. Jamnadass, and S. Anthony. 2009. Agroforestree Database: A Tree Reference and Selection Guide Version 4.0. World Agroforestry Centre, Kenya. http://apps.worldagroforestry.org/treedb2/

Pearson, Richard G., Wilfried Thuiller, Miguel B. Araújo, Enrique Martinez-Meyer, Lluís Brotons, Colin McClean, Lera Miles, Pedro Segurado, Terence P. Dawson, and David C. Lees. 2006. "Model-Based Uncertainty in Species Range Prediction." *Journal of Biogeography* 33 (10): 1704–11. https://doi.org/10.1111/j.1365-2699.2006.01460.x.

Phillips, Sharon B., Viney P. Aneja, Daiwen Kang, and S. Pal Arya. 2006. "Maximum Entropy Modeling of Species Geographic Distributions." *Ecological Modelling* 6 (2–3): 231–52. https://doi.org/10.1016/J.ECOLMODEL.2005.03.026.

Phillips, Steven, Miroslav Dudík, and Robert Schapire. 2004. "A Maximum Entropy Approach to Species Distribution Modeling." *Proceedings, Twenty-First International Conference on Machine Learning, ICML 2004*. Vol. 21. https://doi.org/10.1145/1015330.1015412.

Phillips, Steven J, Robert P Anderson, and Robert E Schapire. 2006. "MaxEnt Software for Modeling Species Niches and Distributions." *Ecological Modelling* 190. http://biodiversityinformatics.amnh.org/open_source/maxent/.

Shi, Xiaodeng, Qun Yin, Ziyang Sang, Zhonglong Zhu, Zhongkui Jia, and Luyi Ma. 2021. "Prediction of Potentially Suitable Areas for the Introduction of Magnolia wufengensis under Climate Change." *Ecological Indicators* 127: 107762. https://doi.org/10.1016/j.ecolind.2021.107762.

Song, Chuangye, Huiming Liu, and Jixi Gao. 2019. "Habitat Preference and Potential Distribution of Magnolia officinalis subsp. officinalis and M. O. subSp. biloba in China." *Nature Conservation* 36 (September): 93–111. https://doi.org/10.3897/natureconservation.36.36171.

Tabuti, John R. S., Collins B. Kukunda, Daniel Kaweesi, and Ossy M. J. Kasilo. 2012. "Herbal Medicine Use in the Districts of Nakapiripirit, Pallisa, Kanungu, and Mukono in Uganda." *Journal of Ethnobiology and Ethnomedicine* 8: 1–15. https://doi.org/10.1186/1746-4269-8-35.

Thuiller, Wilfried, Jordi Vayreda, Joan Pino, Santi Sabate, Sandra Lavorel, and Carles Gracia. 2003. "Large-Scale Environmental Correlates of Forest Tree Distributions in Catalonia (NE Spain)." *Global Ecology and Biogeography* 12 (4): 313–25. https://doi.org/10.1046/j.1466-822X.2003.00033.x.

Tiawoun, M. A. P., M. P. Tshisikhawe, and M Ligavha-Mbelengwa. 2017. "Seed Germination and Seedling Establishment of Securidaca longepedunculata (Polygalaceae)." *Pakistan Journal of Botany* 49 (January): 1009–16.

Tshisikhawe, Milingoni Peter, Olivia Baloyi, M. Ligavha-Mbelengwa, and R. B. Bhat. 2012. "The Population Ecology of Securidaca longepedunculata Fresen in the Nylsvley Nature Reserve, Limpopo Province, South Africa." *Phyton* 81 (January): 107–12. https://doi.org/10.32604/phyton.2012.81.107.

Wang, Fang, Xingzhong Yuan, Yingjun Sun, and Yaohui Liu. 2024. "Species Distribution Modeling Based on MaxEnt to Inform Biodiversity Conservation in the Central Urban Area of Chongqing Municipality." *Ecological Indicators* 158 (December 2023): 111491. https://doi.org/10.1016/j.ecolind.2023.111491.

Welk, Erik, Konstanze Schubert, and Matthias H. Hoffmann. 2002. "Present and Potential Distribution of Invasive Garlic Mustard (Alliaria petiolata) in North America." *Diversity and Distributions* 8 (4): 219–33. https://doi.org/10.1046/J.1472-4642.2002.00144.X.

Wisz, M. S., R. J. Hijmans, J. Li, A. T. Peterson, C. H. Graham, A. Guisan, and NCEAS Predicting Species Distributions Working Group. 2008. "Effects of Sample Size on the Performance of Species Distribution Models." *Diversity and Distributions* 14 (5): 763–73. https://doi.org/10.1111/j.1472-4642.2008.00482.x.

Wyk, B-E. van, and C. Albrecht. 2008. "A Review of the Taxonomy, Ethnobotany, Chemistry and Pharmacology of Sutherlandia frutescens (Fabaceae)." *Journal of Ethnopharmacology* 119 (3): 620–29. https://doi.org/10.1016/j.jep.2008.08.003.

Zulu, Donald, Blackson L. K. Thokozani, Gudeta W. Sileshi, Zewge Teklehaimanot, Dominic S. B. Gondwe, Viswanbharen Sarasan, and Philip C. Stevenson. 2011. "Propagation of the African Medicinal and Pesticidal Plant, Securidaca longepedunculata." *African Journal of Biotechnology* 10 (32): 5988–92.

Chapter 6

Taxonomy of the genus *Securidaca* L.

Anywar Godwin U,[1] Wanyama James,[1] and Douglas Ongeri Ochora[2]

[1]Department of Plant Sciences, Microbiology and Biotechnology, College of Natural Sciences, Makerere University, Kampala, Uganda
Ethnopharmacology and Zoopharmacognosy, Bernhard Nocht Institute for Tropical Medicine, Hamburg, Germany
[2]Department of Biological Sciences, School of Pure and Applied Sciences, Kisii University, Kisii, Kenya

6.1 INTRODUCTION

Research on the taxonomy of members of the genus *Securidaca* has emerged as a critical area of inquiry due to its broad biogeographical distribution and ecological significance across tropical regions of the Americas, Africa, and Asia (Bojase and Ombito 2024). *Securidaca* L. is a pantropical genus of the Polygalaceae family (Aymard-Corredor 2022). Different sources and authors put the number of species in this genus at between 66 and 80 species (Smith et al. 2004; Aymard-Corredor 2022). However, the Plants of the World Online (POWO) database puts the number at 59 accepted species (POWO, 2025). The native range of this genus is tropical and subtropical (POWO, 2025). The largest diversity of *Securicada* spp. is found in the Neotropics (Aymard-Corredor 2023), and a few species are found in Africa and Southeast Asia (Aymard-Corredor 2022).

Members of the genus are characterized by their distinctive winged fruits and zygomorphic flowers. Species of *Securidaca* exhibit considerable morphological diversity (Miller 1971). Within Africa, early taxonomic work on *Securidaca* was notably advanced by Robyns (1927), who contributed to the classification of key species, such as *S. longepedunculata* Fresen and *S. welwitschii* Oliv., and also proposed intraspecific varieties within *S. longepedunculata*. This work laid the foundation for subsequent regional floristic studies and herbarium records that have refined species boundaries and distributional data. More recent taxonomic revisions have used both morphological and molecular data to clarify relationships within the genus, confirming monophyly and supporting its placement within the Polygalaceae family (Pastore and Abbott 2012; Pastore et al. 2019; Forest et al. 2007).

The genus *Securidaca* was first described by Linnaeus in 1754, based on specimens originating from the Americas. It was defined by features such as a keel that may or may not have a small fringed crest, eight stamens with filaments partially fused at the base, and a winged fruit composed of

DOI: 10.1201/9781003544463-6

two locules. The earliest recorded African specimen was collected in 1837 by the German explorer and zoologist Dr W. P. E. Rüppell. Later, in 1842, Hochstetter reassigned the African specimens to a new genus, *Lophostylis* Hochst. He delimited the new genus by the emarginate and bifid keel, the unilocular anthers, and an expanded style which eventually forms the ridge of the samaroid fruit (Johnson 1987).

The taxonomy of the genus *Securidaca* (family Polygalaceae) presents several challenges due to its wide morphological variability, broad geographic distribution, and limited comprehensive phylogenetic studies. Species within the genus exhibit considerable overlap in key diagnostic features, such as leaf shape, floral morphology, and fruit structure, which complicates species delimitation and identification (Bente Eriksen 1993; Paiva 1998). Additionally, many species of *Securidaca* occur in tropical and subtropical regions with underexplored floras, leading to incomplete herbarium representation and potential misidentification. Molecular phylogenetic work on the genus is still in early stages, limiting insights into evolutionary relationships among species and with closely related genera (Forest et al. 2007). The reliance on traditional morphological taxonomy without integration of genetic data has further hindered a clear understanding of species boundaries and intraspecific variation (Yang et al. 2022). These challenges underscore the need for integrative taxonomic approaches combining morphological, molecular, and ecological data to resolve the complex taxonomy of *Securidaca*.

6.2 METHODS

In this chapter, a systematic approach was used to examine the taxonomy of the genus *Securidaca* (Polygalaceae). Relevant literature was identified through comprehensive searches of scientific databases including Web of Science, Scopus, Google Scholar, and JSTOR. We used search terms such as "*Securidaca* taxonomy," "Polygalaceae classification," "morphological variation in *Securidaca*," and "molecular phylogeny of *Securidaca*." Publications from peer-reviewed journals, botanical monographs, and floras provided taxonomic revisions, morphological descriptions, phylogenetic analyses, and regional floristic treatments of *Securidaca*. Data were extracted and synthesized to trace historical changes in classification, identify taxonomic challenges, and evaluate recent advances, particularly those integrating molecular techniques. The review aimed to highlight gaps in knowledge and propose directions for future taxonomic research within the genus.

6.3 DISCUSSION

The genus *Securidaca* includes woody lianas and herbaceous vines or lianas, rarely shrubs or small trees (Smith et al. 2004; Johnson 1987; B Eriksen and Persson 2007). *Securidaca* is absent in Australia (Aymard-Corredor, 2022). *Securidaca* in the Neotropics is most diverse throughout the Amazon (Marques 1996) and Andes bioregions, where the species are found in lowland vegetation and montane forests (Aymard-Corredor, 2022).

Several species are found on rocky slopes and oligotrophic soils derived from the Precambrian crystalline basement of the Guayana Shield, and drained by black water rivers (e.g., *S. savannarum* Wurdack) and on white sandy shrubby savannas and Caatinga forests (Aymard et al., 2004). The remaining species are found in more particular habitats, such as foothills in the low-medium to high altitude (500–2,500 m) mountains of the Andes (e.g., *S. leiocarpa* S.F. Blake, *S. planchoniana* Killip & Dugand), along the Pacific Coast of Colombia and Ecuador (e.g., *S. calophylla* (Poepp.) S.F. Blake) and dry forests (e.g., *S. scandens* Jacq.) (Aymard-Corredor, 2022).

Polygalaceae is monophyletic and is currently divided into four tribes: Carpolobieae B. Eriksen, Diclidanthereae Reveal, Polygaleae Chodat, and Xanthophylleae Chodat (Mota et al. 2019). *Securidaca* is included in the tribe Polygaleae and is characterized by a three-merous corolla (rarely also with two rudimentary lateral petals) and a bicarpellate gynoceium with one ovule per fertile locule (B Eriksen and Persson 2007). *Securidaca* is further distinguished by the keel with a crest well-developed or obsolete, a one-seeded samara, and an unappendaged seed (Aymard and Campbell 2007).

To date, *Securidaca* has not been the subject of a comprehensive monographic treatment, and no formal infrageneric classification has been proposed (Aymard-Corredor, 2022). Early contributions on this genus were made by some early researchers such as Sprengel (1826). Some researchers such as Marques have identified two taxonomically useful inflorescence characters that separate species, and these two inflorescence characters are as follows: flowers solitary or in short racemes 0.5–2.5 cm long, with 5–15 (–20) flowers; and inflorescences racemes or panicles, 2.5–45 cm long, with more than 15 flowers (Marques 1996).

The genus *Securidaca* has been included in some flora such as the flora of Peru (Macbride 1950), flora of Panama (Woodson et al. 1969), flora of Brazil (Marques, 1996), flora of the Venezuelan Guayana (Aymard et al., 2004), and Manual de Plantas de Costa Rica (Morales 2014). Numerous studies have been done involving herbarium research for an update of Polygalaceae in the Nuevo Catálogo de la Flora de Venezuela (Aymard & Campbell, 2007) and Catálogo de Plantas y Líquenes de Colombia (Gradstein and Uribe 2016).

In one study, specimens of a morphologically distinct species of *Securidaca* were discovered after being collected on the upper Rio Negro southwestern Orinoco regions (Aymard-Corredor, 2022). In a geographical and taxonomical context, *Securidaca amazonica* Chodat is treated as distinct from *S. rivinifolia* A. St. Hil. and *S. volubilis* L., while *S. densiflora* Linden & Planchon is considered a synonym of *S. pubescens* DC., and

S. dasycarpa Turcz. as a synonym of *S. diversifolia* (L.) S. F. Blake (Aymard-Corredor, 2022).

Securidaca was established in 1754 by Linnaeus and was based on material from America. The genus was characterized by its keel with or without a small fringed crest, its eight stamens with filaments fused toward the base, and its two- locular, winged fruit (Johnson 1987). The first African collection of the genus was made by the German traveller and zoologist Dr W.P.E. Ruppel in 1837 (Johnson, 1987), and the genus was then placed the African material in a new genus *Lophostyis* Hochst (Hochstetter 1842). He delimited the new genus by the emarginate and bifid keel, the unilocular anthers, and an expanded style which eventually forms the ridge of the samaroid fruit (Johnson, 1987). Based on the shape of the stigma, Robyns (1927) placed the African species in *Securidaca* in section Cristatae (Robyns, 1927) and other authors also contributed toward the knowledge of this taxon (Oliver 1868). In Africa, some species of the genus *Securidaca* have been studied, such as *S. welwitschii* and *S. longepedunculata* (Robyns, 1927).

For species in the genus *Securidaca*, three (sometimes five) petal lobes are present and when there are five, the two lateral ones are abortive and adnate to the base of the staminal sheath (Johnson, 1987). The two upper petals are adnate to the base of the staminal sheath; the median petal, devoid of a fringed crest, is modified into a keel that encloses the stamens; the keel is adnate to the base of the staminal sheath; the eight stamens are partially fused to form a tapering sheath around the ovary (Johnson, 1987). The dorsal side of the sheath is split and pilose; the two-celled anthers are basifixed and dehisce introrsely by an oblique triangular pore; the superior ovary of two united carpels contains a single pendulous ovule in each locule and placentation is axile (Johnson, 1987).

In the mature fruit placentation appears to be marginal and this is the result of the fact that the dorsal carpel is abortive; the wing of the samaroid fruit develops from the median portion of the ovary wall (Johnson, 1987) and the pollen grains are polycolporate, as is characteristic of the family (Erdtman 1952). The grains are isopolar and radiosymmetrical with a diameter of ca. 30 µm (Johnson, 1987). The scanning electron micrographs of pollen of *S. longepedunculata* and S. *welwitschii* reveal that the pollen grains have 10–11 pores that are located in the colpi in an equatorial plane (Coetzee and Robbertse 1985). In terms of diagnostic features, the genus *Securidaca* can be separated from other Polygalaceae on the combination of one locular samaroid fruit, eight stamens, and three petals (Johnson, 1987).

6.3.1 Taxonomic overview of *Securidaca*

The genus is characterized by significant morphological diversity, which has led to taxonomic challenges and frequent revisions (Bojase and Ombito 2024). For example, in Africa, only two species are considered endemic, and their identification relies on detailed morphological keys due to their variation and overlapping features (Johnson 1987). In the Brazilian Amazon, new species such as *S. aurea* and *S. marajoara* have been described recently, highlighting ongoing discoveries and the need for careful morphological comparison with closely related species (dos Anjos, de et al. 2024; Silva da Costa, 2013).

The taxonomy of *Securidaca* is complicated by incomplete original species descriptions and the lack of comprehensive material, such as mature fruits, in herbarium collections. This has resulted in confusion, with some species being lumped together under the same name or, conversely, a single species being described under multiple names. As early as the 1940s, Oort noted that detailed study of type specimens and broader sampling are necessary to clarify species boundaries (Oort 1940).

6.3.2 Taxonomy of notable *Securidaca* species

The genus *Securidaca* is widely distributed in tropical and subtropical regions (Figure 6.1), especially in Africa and South America. Some notable species and their distribution include *S. inappendiculata* and *S. longepedunculata*. These are two of the most studied species, especially for their medicinal properties in Africa and Asia (Bojase and Ombito 2024). *S. longepedunculata* is one of the most widely recognized species in the genus. It is distinguished by its long flower peduncles, purple zygomorphic flowers, and characteristic

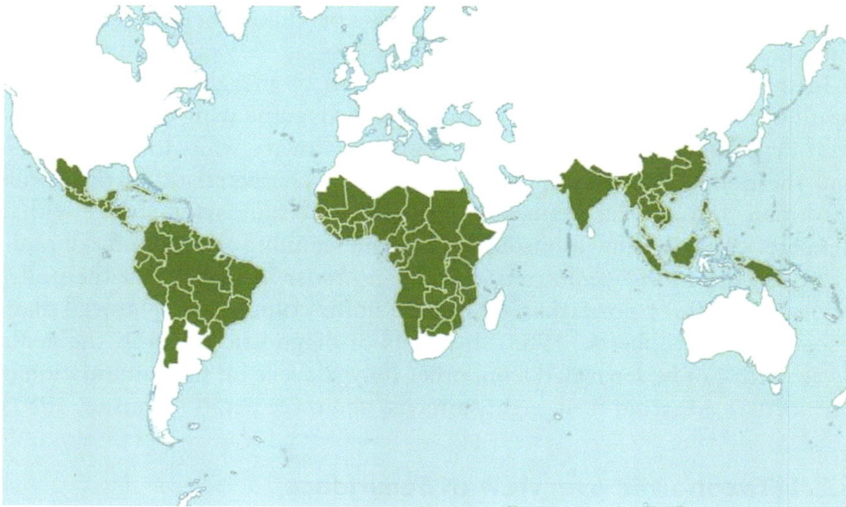

Figure 6.1 Global distribution of the genus *Securidaca*. (Adapted from the Kew Plants of the World Online database [POWO, 2025]. https://powo.science.kew.org/taxon/urn:lsid:ipni.org:names:330544-2 [accessed 18/08/2025].)

samaroid fruits. This species has traditionally been used across sub-Saharan Africa for treating fever, headaches, and snakebites (Neuwinger 2000). Morphologically, it displays considerable intraspecific variation, leading to local misidentification with closely related taxa. However, recent taxonomic treatments have reaffirmed its distinctiveness based on consistent floral and fruit characteristics (Schmelzer and Gurib-Fakim 2008). *S. longepedunculata* is the only species in Polygalaceae family known to have parasyncolpate pollen (Banks et al. 2008).

Recent discoveries in the Amazon, such as *S. aurea* and *S. marajoara*, have shown that the genus continues to expand, with new species differentiated by unique morphological traits like samara length, indumentum, bract persistence, and leaf shape (dos Anjos, and Pastore 2024; Silva da Costa et al., 2013).

Other notable species include *Securidaca diversifolia* (L.) S.F. Blake and *Securidaca pubescens* DC., which have been subjects of taxonomic, ethnobotanical, and pharmacological interest due to their distinctive morphological features and medicinal uses, and *Securidaca welwitschii* Oliv., whose native range is tropical Africa. It is a liana and grows primarily in the seasonally dry tropical biome (POWO, 2025).

***Securidaca longepedunculata* Fresen.** It is a shrub (rarely small tree) 3–6 m high; coloured, branches spreading; shoots spreading, rigid, densely pubescent, becoming spiny, spines bearing leaves, inflorescences. Its leaves with petiole 1.5–3 mm long, blade oblong or elliptic, 1.5–4 (–5) cm by 0.5–1.5 cm and flowers with petals pink to violet-purple while the fruit has a ventral margin of wing much inflected near base. It has been reported in East Africa, including Kenya (Ochora et al., 2022, 2021), Uganda (Schultz et al., 2021), and Tanzania (Joseph et al., 2006); Central Africa, including Angola, Zimbabwe, and Mozambique; and Southern Africa, including Namibia, Zambia, and South Africa (Chinsembu, 2015) in savanna and open forest to 1700 malt and also in stony, sandy, or clayey soil (Johnson, 1987).

***Securidaca welwitschii* Oliv.** This is the second plant species of the genus *Securidaca* which has been reported in Africa. *S. welwitschii* is a large shrub, or often vine, 10–15 m high; stem up to 8 cm in diameter; branches smooth and glabrous, petioles, peduncles and pedicels pubescent and its leaves have petioles 5 mm long; blade broadly elliptical, 5–15 by 2–5.5 cm, apex acuminate, base rounded or rarely subcordinate, thinly coriaceous, glabrous and shiny above, glabrous below (Johnson, 1987).

Furthermore, it has an inflorescence axillary or terminal racemes or simply branched panicles, 2–10(20) cm long, with numerous flowers and the flowers have pedicel 7–10 mm; bracts deciduous; anterior and posterior sepals subequal, ovate or suborbicular, 2–4 mm by 2–4 mm, with wing suborbicular, 5–8 mm in diameter; petals white, yellow or rose, upper petals narrowly obovate, 4.5–6 mm by 2–3 mm, the keel 7–9 mm by 5 mm, stamens with filaments fused into split sheath 3–8 mm long, edge of sheath pilose, free segments 4–6 mm long with second, imperfect wing, seed 6–7 by 5 mm (Johnson, 1987). *S. welwitschii* is distributed in western and central Africa, including Angola and Zambia, and is often a scrambler in savanna forest, and it is also from near sea level to ca. 1900 m altitude (Johnson, 1987). From field notes on herbarium sheets, it appears as if *S. welwitschii* prefers the wetter parts in tropical Africa (Johnson, 1987).

6.4 CONCLUSION

The taxonomy of the genus *Securidaca* remains a complex and evolving field, shaped by both historical morphological classifications and more recent molecular phylogenetic insights that are shaped by ongoing discoveries. Continued research and comprehensive specimen analysis are essential for resolving taxonomic ambiguities and understanding the full diversity of this genus. While early taxonomic treatments relied heavily on variable morphological traits, often leading to ambiguities and inconsistent species delimitation, recent advances in molecular systematics have begun to clarify phylogenetic relationships within the genus and its placement within the Polygalaceae family. However, significant challenges persist, including limited genetic data for many species, underrepresentation of tropical taxa, and unresolved synonymies. This chapter underscores the need for integrative taxonomic approaches that combine classical morphology with molecular, ecological, and biogeographical data. Such efforts will not only improve the resolution of *Securidaca* taxonomy but also contribute to conservation strategies and the sustainable use of species with ethnobotanical importance.

REFERENCES

Aymard-Corredor, Gerardo A. 2022. "Novelties in Securidaca (Polygalaceae, Polygaleae) for the Floras of Colombia and Venezuela." *Harvard Papers in Botany* 27 (2): 143–55. 2023. "An Updated Synopsis and Key to the Species Monnina in the Venezuelan Flora (Polygalaceae, Polygaleae), Including Two New Records." *Harvard Papers in Botany* 28 (1): 15–20.

Aymard, Gerardo A., and Lisa M. Campbell. 2007. "A New Species of *Securidaca* (Polygalaceae) from Sandstone Outcrops in the Venezuelan Andes." *Brittonia* 59: 328–33.

Aymard, G. A., P. E. Berry, and B. Eriksen. 2004. Polygalaceae. *Flora of the Venezuelan Guayana* 8: 316–347.

Banks, H., B. B. Klitgaard, F. Claxton, F. F., and P. R. Crane. 2008. "Pollen Morphology of the Family Polygalaceae (Fabales)." *Botanical Journal of the Linnean Society* 156 (2): 253–89. https://doi.org/10.1111/j.1095-8339.2007.00723.x.

Bojase, Gomotsang, and Ombito Japheth O. 2024. "Xanthones from *Securidaca inappendiculata* and *Securidaca longepedunculata*: Chemistry and Biological Activity." *Natural Products Journal* 14 (1): 89–104.

Chinsembu, K. C. 2015. Plants as antimalarial agents in Sub-Saharan Africa. *Acta Tropica* 152: 32–48. https://doi.org/10.1016/j.actatropica.2015.08.009.

Coetzee, Hester, and P J Robbertse. 1985. "Pollen and Tapetal Development in *Securidaca longepedunculata.*" *South African Journal of Botany* 51 (2): 111–24.

dos Anjos, C. B., de Aguiar D., and J. F. B. Pastore. 2024. "*Securidaca aurea* (Polygalaceae, Polygaleae), a New Species from the Brazilian Amazon." *Kew Bulletin* 79 (4): 955–961. https://doi.org/10.1007/s12225-024-10211-6.

Erdtman, G. 1952. Pollen Morphology and Plant Taxonomy. *Angiosperms.* Almqvist & Wiksell, Stockholm, Sweden.

Eriksen, Bente. 1993. "Phylogeny of The Polygalaceae and Its Taxonomic Implications." *Plant Systematics and Evolution* 186 (1): 33–55. https://doi.org/10.1007/BF00937712.

Eriksen, B., and C. Persson. 2007. "Polygalaceae: Polygalaceae Hoffmanns. & Link, Fl. Portug. 1: 62 (1809), Nom. Cons." In *Flowering Plants· Eudicots: Berberidopsidales, Buxales, Crossosomatales, Fabales Pp, Geraniales, Gunnerales, Myrtales Pp, Proteales, Saxifragales, Vitales, Zygophyllales, Clusiaceae Alliance, Passifloraceae Alliance, Dilleniaceae, Huaceae, Picramniaceae*, 345–63. Springer.

Forest, Félix, Mark W. Chase, Claes Persson, Peter R. Crane, and Julie A. Hawkins. 2007. "The Role of Biotic and Abiotic Factors in Evolution of Ant Dispersal in the Milkwort Family (Polygalaceae)." *Evolution* 61 (7): 1675–94. https://doi.org/10.1111/j.1558-5646.2007.00138.x.

Gradstein, S. R., and J. Uribe. 2016. "Marchantyophyta (Hepáticas)." In *Catálogo de Plantas y Líquenes de Colombia.* Edited by R. Bernal, S. R. Gradstein and M. Celis. Bogotá: Instituto de Ciencias Naturales, Universidad Nacional de Colombia.

Hochstetter, C. F. 1842. "Nova Genera Plantarum Africae Tum Australis Tum Tropicae Borealis Proponit et Describit." *Continuatio. Flora, oder Allgemeine Botanische Zeitung* 25: 225–240.

Joseph, C. C., Moshi, M. J., Sempombe, J., and Nkunya, M. H. H. 2006. "(4-Methoxy-Benzo[1,3]Dioxol-5-yl)-Phenylmethanone: An Antibacterial Benzophenone from *Securidaca longepedunculata.*" *African Journal of Traditional, Complementary and Alternative Medicines*, 3(3): 80–86. https://doi.org/10.1109/TEMU.2014.6917738

Johnson, C. T. 1987. "Taxonomy of the African Species of *Securidaca* (Polygalaceae)." *South African Journal of Botany* 53 (1): 5–11. https://doi.org/10.1016/S0254-6299(16)31465-X.

Macbride, J. F. 1950. "Polygalaceae, Flora of Peru." *Publication Field Museum of Natural History, Botanical Series* 13 (3/3): 891–950.

Marques, M. C. M. 1996. "*Securidaca* L.(Polygalaceae) Do Brasil." *Arquivos Do Jardim Botânico Do Rio de Janeiro* 34 (1): 7–144.

Miller, Norton G. 1971. "The Polygalaceae in the Southeastern United States." *Journal of the Arnold Arboretum* 52 (2): 267–84. www.jstor.org/stable/43781726.

Morales, J. 2014. "Polygalaceae." In: Manual de Plantas de Costa Rica. Vol. VII., B.E. Hammel, M.H. Grayum, C. Herrera and N. Zamora (Eds.). *Monographs in Systematic Botany from the Missouri Botanical Garden* 129: 342–57.

Mota, Michelle, J. Richard Abbott, Roberto M. Salas, Kurt M. Neubig, and José Floriano B. Pastore. 2019. "Three Lonely Argentines: Toward a New Generic Delimitation in Polygalaceae." *Taxon* 68 (3): 522–36.

Neuwinger, H. D. 2000. *African Traditional Medicine. A Dictionary of Plant Use and Applications. African Traditional Medicine. A Dictionary of Plant Use and Applications*. Stuttgart: Medpharm Scientific Publishers.

Ochora, Douglas O., Esezah Kakudidi, Jane Namukobe, Matthias Heydenreich, Paolo Coghi, Li Jun Yang, Edwin W. Mwakio, et al. 2021. "A New Benzophenone, and the Antiplasmodial Activities of the Constituents of *Securidaca longipedunculata* Fresen (Polygalaceae)." *Natural Product Research* 36 (11): 2758–66. doi:10.1080/14786419.2021.1925272.

Ochora, Douglas O., Esezah K. Kakudidi, Jane Namukobe, Perpetua Ipulet, Dancan M. Wakoli, Winnie Okore, Edwin W. Mwakio, Redempthah A. Yeda, Agnes C. Cheruiyot, Dennis W. Juma, et al. 2022. "Synergism in Antiplasmodial Activities of Artemether and Lumefantrine in Combination with *Securidaca longipedunculata* Fresen (Polygalaceae)" *Plants* 11, no. 1: 47. https://doi.org/10.3390/plants11010047

Oliver, Daniel. 1868. *Flora of Tropical Africa*. Vol. 1. Ashford: L. Reeve and Company.

Bojase, G., & Ombito, J. O. (2024). Xanthones from Securidaca inappendiculata and Securidaca longepedunculata: Chemistry and Biological Activity. *The Natural Products Journal, 14*(1), 89–104.

Oort, Arend Joan Petrus. 1940. "Critical Remarks on the Suriname Species of the Genus Securidaca (Polygalaceae)." *Recueil Des Travaux Botaniques Néerlandais* 36 (2): 677–85.

Paiva, Jorge Américo Rodrigues. 1998. *Polygalarum Africanarum et Madagascariensium Prodromus*. Madrid : Real Jardín Botánico. www.biodiversitylibrary.org/bibliography/74345

Pastore, J. F. B., and J. R. Abbott. 2012. "Taxonomic Notes and New Combinations for Asemeia (Polygalaceae)." *Kew Bulletin* 67 (4): 801–13. https://doi.org/10.1007/s12225-012-9397-x.

Pastore, José Floriano Barêa, J. Richard Abbott, Kurt M. Neubig, Cassio Van Den Berg, Michelle Christine De Almeida Mota, Andressa Cabral, and W. Mark Whitten. 2019. "Phylogeny and Biogeography of Polygala (Polygalaceae)." *TAXON* 68 (4): 673–91. https://doi.org/10.1002/tax.12119.

Plants of the World Online (2025). Royal Botanic Gardens, Kew. https://powo.science.kew.org; https://powo.science.kew.org/taxon/urn:lsid:ipni.org:names:330544-2 (accessed 03/08/2025).

Robyns, Walter. 1927. "Les Espèces Du Genre Securidaca Du Congo Belge." *Bulletin de La Société Royale de Botanique de Belgique/Bulletin van de Koninklijke Belgische Botanische Vereniging* 60 (Fasc. 1): 88–93.

Schmelzer, G. H., and A. Gurib-Fakim. 2008. *Plant Resources of Tropical Africa 11(1)-Medicinal Plants 1*. Edited by Netherlands. 791pp PROTA Foundation, / Backhuys Publishers, /CTA, Wageningen. Vol. 11. Leiden, Wageningen, Netherlands: Prota.

Schultz, F., O. F. Osuji, B. Wack, G. Anywar, and L. A. Garbe. 2021. "Antiinflammatory Medicinal Plants from the Ugandan Greater Mpigi Region Act as Potent Inhibitors in the COX-2/PGH2 Pathway." *Plants*, 10(351): 1–28.

Silva da Costa, Christiane, Ana Cristina Andrade De Aguiar-Dias, and André Olmos Simões. 2013. "*Securidaca marajoara* (Polygalaceae), a New Species from the Brazilian Amazon." *Phytotaxa* 137 (1): 53.

Smith, Nathan, Scott A Mori, Andrew Henderson, Dennis Wm Stevenson, and Scott V Heald. 2004. Flowering Plants of the Neotropics. Princeton: Princeton University Press.

Sprengel, Kurt Polycarp Joachim. 1826. Systema vegetabilium. Editio decima sexta / curante Curtio Sprengel, equite stellae polaris et aquilae rubrae, prof. med. et rei herb. in Univers. Hal. Volumen III. Classis 16–23.

Woodson, Robert E., Robert W. Schery, Walter H. Lewis, and Olga Herrera-MacBryde. 1969. "Flora of Panama. Part VI. Family 96. Polygalaceae." *Annals of the Missouri Botanical Garden* 56 (1): 9–28.

Yang, Bing, Zhenxin Zhang, Cai-Qing Yang, Ying Wang, Michael C Orr, Hongbin Wang, and Ai-Bing Zhang. 2022. "Identification of Species by Combining Molecular and Morphological Data Using Convolutional Neural Networks." *Systematic Biology* 71 (3): 690–705. https://doi.org/10.1093/sysbio/syab076.

Chapter 7

The antimicrobial potentials of plants in the genus *Securidaca*

Abiola Grace Femi-Adepoju[1,*] *and Adeyinka Olufemi Adepoju*[2]
[1]College of Science and Allied Health, African Methodist Episcopal University, Monrovia, Liberia
[2]School of Graduate Studies, African Methodist Episcopal University, Monrovia, Liberia

7.1 INTRODUCTION

Securidaca, a genus of plants native to tropical and subtropical regions, is known for its medicinal properties. The most prominent species with antimicrobial importance is *Securidaca longepedunculata* used in African and Asian medicine (Ombito & Bojase, 2023). These plants, often shrubs or small trees, are widely used in folk medicine to treat infections, inflammatory conditions, and ailments like fever, cough, and snakebites. Certain species of *Securidaca* are native to tropical subregions and alkaloids, glycosides, flavonoids, tannins, and saponins have all been reported in them. Extracts from members of this genus have demonstrated efficacy against a range of microorganisms, such as viruses, bacteria, fungus, and protozoa. *S. longepedunculata* extracts may be useful in the treatment of trypanosomiasis and malaria (Adiele et al., 2013).

According to Schultz et al. (2021), *Securidaca* spp. are among the most understudied plant species in Africa. There have been earlier reports about the therapeutic usefulness of the plants in this genus from ethnobotanical studies (Kumiko et al., 2020; Awoke et al., 2024). Members of the *Securidaca* genus have formed a part of the most culturally important medicinal plant species that have shown therapeutic usage for several types of diseases including those caused by varieties of microbes as reported by Jima et al. (2021), Novotna et al. (2020), and Mongalo et al. (2015). *S. longepedunculata*, family Polygalaceae, also known as violet tree, fiber tree, or Muteya/Mzingi in Swahili, is a native African medicinal plant. It is commonly found in the Savannah and Sahel regions of sub-Saharan Africa. The plant exhibits a wide range of medicinal, pharmacological, and therapeutic qualities and has been utilized for a number of purposes in African traditional medicine (Anjarwalla et al., 2015; Mahomoodally, 2013; Maroyi, 2013, Awoke et al., 2024). *S. longepedunculata* possess antimicrobial, antioxidant, antiparasitic,

*Correspondence: adeyinka.adepoju.phd@gmail.com

 DOI: 10.1201/9781003544463-7

antidiabetic, anti-inflammatory, antimalarial, insecticidal, molluscicidal, pesticidal, and anticonvulsant properties (Makangara et al., 2024).

7.2 METHODOLOGY

Data utilized were from databases such as PubMed, Scopus, Web of Science, Science Direct, Google Scholar, Wiley Online Library, and university repositories. Published articles between the period 1993–2024 were included to encompass the earliest and most current studies. Inclusion criteria were original and review research articles published in English language. Exclusion criteria were articles not specifically on *Securidaca* species and editorials, letters, and opinion pieces (used only for background).

7.3 ANTIMICROBIAL COMPOUNDS REPORTED IN MEMBERS OF THE GENUS *SECURIDACA*

One of the most extensively investigated species of the *Securidaca* genus, *S. longepedunculata*, contains a variety of compounds, including sterol glycosides, phenols, tannins, steroids, phytosterols, alkaloids, flavonoids, phenols, and tannins (Table 7.1). These substances are present in the stem bark, roots, and leaves of the plants. Studies have noted that these compounds are effective against microbial illnesses such as gonorrhea and urinary tract infections (UTIs). The roots and stem bark contain saponins, which have been shown to be effective against UTIs. There has also been evidence of the presence of special alkaloids in the plant species, including indolizidine alkaloids (a derivative of securinine). In the root and leaves of *S. inappendiculata*, xanthone, securidanes, sterols, and sterol glycosides have been identified with documented biochemical activity. *S. longepedunculata* and *S. inappendiculata* have a wide range of potential antimicrobial chemicals that have been are beneficial in treating a variety of infections.

7.4 ANTIMICROBIAL PROPERTIES OF *SECURIDACA* SPECIES

Plant species in the genus *Securidaca* have shown various antimicrobial activities. *S. longepedunculata* is used to treat various ailments (Ngulde et al., 2019). According to Semenya and Potgieter (2013), traditional healers in Ethiopia frequently use it to treat fungal infections and sexually transmitted diseases caused by both bacteria and fungi.

7.4.1 Antibacterial properties

Plants have been used historically for their medicinal properties, with records dating back to ancient civilizations. They have been a cornerstone

Table 7.1 Earlier reported antimicrobial potentials of *Securidaca longepedunculata*

References and Conc. Used	*Test Organisms*	*Method*	*Plant Part*	*Solvents*	*Compound*
1. Pallant & Steenkamp (2008) 1 mg/mL	*M. smegmatis* *H. influenzae* *K. pneumoniae* *S. aureus* *S. pneumoniae*	Agar disc diffusion	Root bark	Methanol and H_2O	Saponins, alkaloids, flavonoids, phenols, triterpenes
2. Makinia et al. (2020) 0.01 mg/mL	*N. gonorrhea*	Agar disc diffusion	Root and bark	Ethanol	Saponins
3. Junaid et al. (2013) 200 mg/mL	*E. coli* *S. typhi* *Shigella* spp.	Agar disc diffusion	Root and leaves	H_2O, Ethanol	Tannins, saponins, cardiac glycosides, flavonoids, alkaloids
4. Namadina et al. (2020) 62.5–500 mg/mL	*E. coli* *S. aureus* *K. pneumoniae*	Agar well diffusion	Roots	H_2O, Methanol	Saponins, alkaloids, flavonoids, phenols, terpenes, glycosides, tannins, cardiac glycosides, triterpenes
5. Ngonda et al. (2012) 0.3125–10 mg/mL	*S. aureus* *P. aeruginosa* *E. coli* *C. albicans*	Agar well diffusion	Roots	Ethanol, Acetone, H_2O	Flavonoids, alkaloids, saponins, phenols, phytosterols, tannins, terpenes
6. Adebayo & Osman (2012) 200–100 mg/mL	*E. coli* *S. typhi*	Agar well diffusion	Root bark	Ethanol, H_2O, Pet. Ether, Chloroform, Methanol	Tannins, alkaloids, coumarins, flavonoids, saponins
7. Tikisa et al. (2019) 200 mg/mL	*S. aureus* *E. coli* *P. aeruginosa* *B. subtilis*	Disc diffusion	Root bark	Hexane, Chloroform Acetone, Methanol	Two benzoic acid derivatives (1 and 2) and a xanthone (3)

References and Conc. Used	*Test Organisms*	*Method*	*Plant Part*	*Solvents*	*Compound*
8. Abubakara et al. (2018) 20 mg/mL	*S. aureus* *E. coli* *S. typhi* *P. aeruginosa* *S. dysenteriae* *K. pneumoniae*	Agar well diffusion	Stem bark	H_2O	Saponins, alkaloids, flavonoids, phenols, terpenes, glycosides, tannins, triterpenes
9. Melaku et al. (2022) 0.5–1 mg/mL	*S. aureus*	Disc diffusion	Stem bark	*N*-Hex Fraction	1,7-dihydroxy-4-methoxyxanthone
10. Apak & Olila (2006) 1 mg/mL	*S. aureus* *E. coli*	Agar well diffusion	Root bark	H_2O, Methanol	—
11. Odongo et al. (2022) 1 mg/mL	*S. aureus*	Disc diffusion	Root bark	H_2O	—
12. Joseph et al. (2006) 10 mg/disc	*P. aeruginosa* *S. aureus* *A. fumigatus* *A. niger* *Penicillum* sp. *C. albicans*	Disc diffusion	Roots	Pet. Ether, Dichloromethane, Methane	4-methoxy-benzo[1,3] dioxol-5-yl)-phenylmethanone 1,7-dihydroxy-4-methoxyxanthone
13. Datagni et al. (2021) 250 mg/mL	*S. aureus* *C. albicans*	Agar well diffusion	Roots	Hydro-alcohol	Alkaloids, tannins, flavonoids phenols, saponins
14. Junaidu et al. (2014) 300 mg/mL	*A. flavus* *A. niger*	Agar incorporation	Root bark	Methanol	Cardiac glycosides, flavonoids, saponins, alkaloids, saponin glycoside
15. Muhammad (2021) 150 mg/mL	*Trichophyton rubrum* *Trichophyton schoenleini* *Microsporum audouinii*	Disc diffusion	Leaves	*N*-hexane, Ethanol, Acetone, H_2O	Phytosterols, tannins, flavonoids, alkaloids, saponins

(*continued*)

Table 7.1 (Coninued)

References and Conc. Used	*Test Organisms*	*Method*	*Plant Part*	*Solvents*	*Compound*
16. Alitonou et al. (2012) 12.79 and 0.40 mg/mL	*E. coli* *C. albicans*	Well microdilution / microtiter	Root bark	H_2O	Monoterpene hydrocarbons Oxygenated monoterpenes Sesquiterpene hydrocarbons Oxygenated sesquiterpenes
17. Runyoro et al. (2006) 200 μg/μL	*Candida albicans*	Bioautography agar overlay method	Root bark	Methanol	—
18. Nitiema et al. (2020) 100 mg/mL	*Fusarium solani* *Fusarium maniliforme* *Curvularia lunati*	Disc diffusion	Stem bark	Hydro-ethanol	—
19. Samie & Mashau (2013) 7.5–1.9 mg/mL	*F. verticilliodes* *F. oxysporum* *F. nygamai* *F. proliferatum* *F. graminearum*	Hole plate diffusion	Root	Acetone, *N*-hexane	—
20. Mahmood et al. (1993) 0.32–0.6 μg/mL	HIV-1 III-B SARS-CoV-2 *virus*	Well microdilution / microtiter	Roots	Methanol	Caffeineoylquinic acids
21. Viol et al. (2016) 351.89 ug/mL	Herpes simplex virus 2 (HSV-2)	Neutralization	Roots	Methanol	—
22. Haruna et al. (2013a) 150 and 300 mg/kg 23. Haruna et al. (2013b) 200–100 mg/kg 0.56–0.14 mg/kg	*P. berghei* *T. brucei*	*In vivo*	Roots	Methanol, H_2O	Triterpenes, steroids, alkaloids tannins, glycosides, flavonoids

References and Conc. Used	*Test Organisms*	*Method*	*Plant Part*	*Solvents*	*Compound*
24. Amos et al. (2019) 34, 23, and 11 mg/kg 25. Nguta (2019) 100 mg/kg	*P. berghei*	*In vivo*	Root bark Whole part	Cysteine Protease Inhibitors' Fraction H_2O, Chloroform + Methanol (1:1)	Alkaloids; tannins, steroids, triterpenes, flavonoids, cardiac glycosides, saponins
26. Issa et al. (2018) 2.2 and 2.6 µg/mg 27. Karama et al. (2020) 6.98 and 10.39 µg/mL 28. Bah et al. (2007) 7 ug/mL 29. Ochora et al. (2021) 28.8 and 18.6 µM 30. 19.7 and 14.5 µM	*Plasmodium falciparum*	*In vivo* *Ex vivo*	Leaves Leaves and roots Root bark	Methanol, Chloroform, Flavonoid Fractions, Dichloromethane, Methanol	Flavonoids, phenolic compounds, and alkaloids Benzophenones
31. Atawodi et al. (2003) 0.4 µg/mL 32. Atawodi (2005) 2 and 4 µg/mL	*T brucei* *T. congolense* *T brucei*	WM	Roots Stem bark	Methanol, Pet. Ether	—
33. Tauheed et al. (2016) 3 and 6 mg/mL	*T. brucei*	*Ex vivo*	Stem bark	Crude Methanol	—

in traditional medicine systems worldwide (Ahmad & Wajid, 2013). *S. longepedunculata* is a versatile plant utilized in traditional medicine in African countries. The antibacterial potential was observed against some test organisms namely: *Escherichia coli*, *Salmonella typhi*, *Staphylococcus aureus*, *Shigella dysenteriae*, *Pseudomonas aeruginosa*, and *Klebsiella pneumoniae*. With the highest activity against *S. aureus* (36 mm) and the lowest activity against *S. dysenteriae* (24 mm), the aqueous stem bark extract was shown to be effective against all tested bacterial isolates. *S. longepedunculata* is a useful medication for treating bacterial infections (Abubakara et al., 2018). A study by Junaid et al. (2013) evaluated the antimicrobial activity of aqueous, ethanol, and methanol extracts from the roots and leaves of *S. longepedunculata* against *E. coli*, *Shigella* sp., and *S. typhi*, using the agar gel diffusion method. The results indicated that the aqueous root extract was the most effective, with minimum inhibition concentration (MIC) and minimum bacteriocidal concentration (MBC) values of 50 mg/mL and 200 mg/mL, respectively, particularly against *Shigella* spp. and *S. typhi*. The research also highlighted that leaf extracts and methanolic root extracts showed no antimicrobial activity, suggesting that differences in phytochemical composition among the extracts could account for the varying levels of antimicrobial effectiveness. The findings support the traditional use of the plant in treating diarrhoea and suggest its potential as a source of antimicrobials amid rising drug resistance (Junaid et al., 2013). The methanol root extract of *S. longepedunculata* showed encouraging antibacterial activity against UTI-causing bacteria. *S. aureus* was the next most susceptible bacteria with an inhibition zone of 16 mm and MIC of 31.25 mg/mL, while *E. coli* was found to be the most susceptible bacteria in both the aqueous and methanol extracts of *S. longepedunculata* at 500 mg/mL with inhibition zones of 20 and 16 mm (Namadina et al., 2020). In addition, a study by Mongalo et al. (2015) reported the efficacy of *S. longepedunculata* against human pathogenic bacteria that cause illnesses like tuberculosis, pneumonia, and UTIs.

Pallant and Steenkamp (2008) used the disc diffusion assay to assess the antimicrobial activity of crude methanol and water extracts of *S. longepedunculata* against standard strains and clinical isolates of *Haemophilis influenzae*, *K. pneumoniae*, *S. aureus*, *Streptococcus pneumoniae*, and *Mycobacterium smegmatis*. Significant antibacterial activities were reported against the test organisms at an MIC of $\leq$1 mg/mL.

Luo et al. (2011) tested the *n*-hexane, dichloromethane, ethyl acetate, and 70% ethanol extract of *S. longepedunculata* root bark against *M. smegmatis* ATCC 607 and *M. tuberculosis* H37Rv. The extract was deemed active with MIC of less than 125 µg/mL. The study assessed the cytotoxic impact of *n*-Hexane extracts of *S. longepedunculata* on human macrophages using monocytic THP-1 cells. The results showed significant efficacy against *M. tuberculosis* H37Ra and *M. smegmatis* BCG, with MICs ranging from

15.6 to 62.5 μg/mL. This suggests the potential of *S. longepedunculata* as potential anti-TB drugs.

According to Tikisa et al. (2019), the crude extracts of *S. longepedunculata* root bark demonstrated significant antibacterial activity in an *in vitro* antibacterial assay against four bacterial strains (*S. aureus* ATCC 25923, *B. subtilis* ATCC 6633, *E. coli* ATCC 35218, and *P. aeruginosa* ATCC 27853), while the acetone extract was shown to exhibit the highest activity against *B. subtilis* and *P. aeruginosa*. The root bark of *S. longepedunculata* exhibited bactericidal activity at 0.01 μg/mL of 10% crude extract and a zone of inhibition of ≥ 10 mm against two common isolates of *Neisseria gonorrhoeae*, NO.0296 and WHO-K (Makinia et al., 2020).

Pauw and Eloff (2014) investigated the antibacterial activity of *S. longepedunculata* acetone extract on the growth and development of bacteria strains: *Enterococcus faecalis*, *S. aureus*, *E. coli*, and *P. aeruginosa*. The report was based on the sensitive serial dilution microplate procedure, and it was found to have bactericidal activity against both the Gram-positive and Gram-negative bacteria with an MIC as low as 0.02 mg/mL. At a concentration of 300 mg/mL, the antibacterial activities of *S. longepedunculata* were 21.55 ± 0.65 mm and 13.15 ± 0.15 mm against *E. coli* and *K. pneumonia*, respectively. For *E. coli* and *Pseudomonas*, the lowest MIC and MBC values were 37.5 and 75 mg/mL. *S. longepedunculata* could be a good candidate for use in managing bacteria and related illnesses, although further evaluation is needed to fully validate its usage (Ibrahim, 2021).

Melaku et al. (2022) evaluated the antibacterial activity of the constituents of *S. longepedunculata*, using a novel *in silico* molecular docking analysis of the isolated compounds. Using *n*-hexane, EtOAc, and MeOH, each 2.5 L for 72 hours, the 500 g of stem bark of *S. longepedunculata* was extracted in stages and distinct compounds isolated. Particularly, compound 1,7-dihydroxy-4-methoxyxanthone significantly inhibited the growth of *S. aureus* by 16.06 ± 0.22 mm. The findings suggest that this compound could be regarded as a potential candidate for the development of antimicrobial drugs (Melaku et al., 2022).

Akpemi et al. (2013) conducted a study on the antimicrobial activity of the fractionated root bark extract of *S. longepedunculata*. The *n*-hexane, chloroform, and ethyl acetate fractions were then tested against the Gram-positive bacteria: methicillin resistant-*S. aureus*, *S. pyogenes* and Gram-negative bacteria: *S. typhi*, *K. pneumoniae*, *P. aeruginosa*, and *P. fluorescens*. All six pathogenic bacteria assayed were susceptible to the four fractions. The chloroform fraction had the highest mean zone of inhibition (23 mm), followed by the methanol fraction (21 mm), ethyl acetate fraction (19 mm), and hexane fraction (15.5 mm).

The ethanol, acetone, and water crude extracts of *S. longepedunculata* root have shown activity against *S. aureus*, *P. aeruginosa*, and *E. coli*. When the extract was tested for bioactivity the diffusion assay, the range of minimum inhibitory concentration (MIC) against *P. aeruginosa* and *E. coli* in *n*-butanol fractions was 0.156 to 0.3125 mg/mL (Ngonda et al., 2012).

The Jiangsu Institute of Botany in 1998 also reported that the roots of *S. inappendiculata* are utilized as antibacterial medicines in traditional Chinese medicine. This can be attributed to the bioactive compounds like xanthones, saponins, sugar esters, organic acids, and benzophenones (Yang et al., 2020). Adebayo and Osman (2012) also reported significant antibacterial activity with a zone of inhibition of 15.10 mm and 12 mm by the ethanol extracts of *S. longepedunculata* root bark at a concentration of 100 mg/mL against *E. coli* and *S. typhi*. At double the the concentration (200 mg/mL), the ZI were recorded as 18.20 and 15 mm, respectively.

The aqueous root bark extract of *S. longepedunculata* demonstrated activity against *S. aureus* ATCC 27853 and *S. aureus* ATCC 25923 but showed little or no activity against *E. coli* ATCC 25922 in Uganda (Apak & Olila, 2006). The hydroalcoholic root extract of *S. longepedunculata* demonstrated antibacterial efficacy against *S. aureus* (inhibition zone = 12.03 ± 0.03 mm, MIC = 125 mg/mL, MBC = 250 mg/mL) (Datagni et al., 2021). The study proved that *S. longepedunculata* had antibacterial characteristics and is a potential candidate for the treatment of infections following snakebite. According to Odongo et al. (2022), the root bark extract of *S. longepedunculata* has also shown significant inhibitory activity against *S. aureus*. At 1 g/mL, the aqueous root extract showed an inhibition zone of 12.5 ± 2.2 mm, while streptomycin had an inhibition zone of 18.0 ± 0.1 mm at 25 μg/mL.

Joseph et al. (2006) demonstrated the antibacterial properties of 4-methoxy-benzo[1,3]dioxol-5-yl)-phenylmethanone against *P. aeruginosa* and the antibacterial activity of 1,7-dihydroxy-4-methoxyxanthone against *S. aureus*. These two compounds have been isolated from the root extract of *S. longepedunculata*. Further studies have emphasized that the stem bark extract possesses antibacterial properties against clinical isolates of *E. coli, S. typhi, S. aureus, S. dysenteriae, P. aeruginosa,* and *K. pneumoniae* (Makangara et al., 2024). Various ethnobotanical studies have also reported on the antibacterial efficacy of *S. longepedunculata*. For instance, the antisymptomatic effect of *S. longepedunculata* against *Mycobacterium* infection by Green et al. (2010) and STI by Mongalo and Raletsena (2022).

7.4.2 Antifungal properties

Fungal infections affect approximately 40 million people globally, with skin infections being among the most common dermatological issues (Gupta et al., 2024). The rise in drug-resistant fungal strains and the limited number of effective antifungal drugs exacerbate the challenge of managing these infections.Junaidu et al. (2024) investigated the antifungal properties of root bark extracts and leave extracts from *S. longepedunculata* against two species of *Aspergillus*: *flavus* and *A. niger* using the agar incorporation method. demonstrating high antifungal activity at 300 mg/mL. They also showed the paln to contain cardiac gylcosides, flavonoids, saponins, and alkaloids. Thus, the phytochemical components of *S. longepedunculata* leaves and root bark extracts may serve as an alternative to the current antibiotics in the management of pathogenic *Aspergillus* infections (Junaidu et al., 2014).

Muhammad (2021) evaluated the antifungal activities of the *n*-hexane, ethyl-acetate and residual aqueous leaf extract of *S. longepedunculata* on dermatophytes. The isolates used were *Trichophyton rubrum, T. schoenleinni*, and *Microsporum audouinii*. .The extract demonstrated dose-dependent antifungal activity against dermatophytes, with the highest inhibition observed at 150 mg/mL (16.65 ± 0.35 mm) and the lowest at 7.81 mg/mL. The minimum fungicidal concentration (MFC) further confirmed its fungicidal potential. These findings suggest that the plant material may serve as a promising candidate for the management of dermatophytic infections.

Nitiema et al. (2020) investigated the antifungal properties of *S. longepedunculata* stem bark hydro-ethanolic extract against *Fusarium solani*, *Fusarium moniliforme*, and *Curvularia lunata*, and assessed the percentages of infected rice seed germination and infection. The three fungal strains were examined for *in vitro* antifungal activity over a 15-day period using the poisoned food methodology at various extract concentrations of 0.25, 0.5, and 1%. *In vivo* antifungal activity against rice seedlings infected with these three fungal strains was assessed using the same extract concentrations. The extract of *S. longepedunculata* totally inhibited the mycelial growth of *F. solani* and *F. moniliforme*, exhibiting an antifungal activity. After 5 days of culture, 1% extract suppressed *C. lunata* mycelial development by 84.7%.

Furthermore, Ngonda et al. (2012) confirmed that the *n*-butanol fraction and an acetone root extract of *S. longepedunculata* had MFCs and MICs values of 1.25 and 2.5 mg/mL, respectively, against *C. albicans*. An 80% methanol extract of the root bark of *S. longepedunculata* exhibited a zone of inhibition of < 4 mm against *C. albicans* (Runyoro et al. (2006). In addition, its root bark essential oils had MIC of 0.40 mg/mL against *C. albicans* as published by according to Alitonou et al. (2012). Samie and Mashau (2013) showed that the acetone root extracts exhibited MIC of 3.75 mg/mL against *F. verticillioides* and *F. oxysporum* while hexane root extract exhibited MFC of 3.75 mg/mL against *F. verticillioides, F. nygamai, F. proliferatum*, and *F. graminearum*. Datagni et al. (2021) also researched the efficacy of *S. longepedunculata* hydroalcoholic root extract against *C. albicans* and reported MIC of 62.5 mg/mL, MFC of 125 mg/mL and 12.12 ± 0.07 mm zone of inhibition. *Aspergillus fumigatus* and

A. niger were susceptible to the antifungal effects of 4-methoxy-benzo[1,3] dioxol-5-yl)-phenylmethanone and 1,7-dihydroxy-4-methoxyxanthone (11.0 ± 1.7 and 13.0 ± 1.0), which were isolated from the root bark extract of *S. longepedunculata*. 1,7-dihydroxy-4-methoxyxanthone also demonstrated activity against *Penicillum* sp. (18.0 ± 1.7) and *Aspergillus niger* (13.0 ± 2.3) (Joseph et al., 2006). The essential oil obtained from the root bark extract of *S. longepedunculata* has significant antifungal effects against *Candida albicans* ATCC 14133 with an MIC 0.40 ± 0.02 mg/mL (Alitonou et al., 2012).

7.4.3 Antiviral properties

The original research on the antiviral effect of *Securidaca* species is scantly reported, but some ethnopharmacological evidences are also available. One major antiviral study of *S. longepedunculata* is that of Mahmood et al. (1993). The study investigates the anti-HIV and anti-SIV activity and toxicity of caffeineoylquinic acids, namely 3,4,5-tri-O-caffeoylquinic isolated from the methanol root extract of *S. longepedunculata*, which has irreversible binding with gp120 and inactivates the virus. The compounds have been tested using C8166 cells infected with HIV-1 III-B, acid, 4,5-di-O-caffeoylquinic acid, caffeineic acid, rosmarinic acid, and synapoic acid. The tested compounds had no anti-HIV activity against HIV-1 and HIV-2 . The therapeutic index of the methanol root extract was 3.8, and the reduction factor (RFb) was 103.

Further details of the study were reported by Viol et al. (2016). The cytotoxicity profile of the methanol root extract of *Securidaca longepedunculata* was evaluated using multiple assays. In the brine shrimp lethality test (BSLT), the LC_{50} (lethal concentration causing 50% mortality) was 351.89 ± 35.70 µg/mL (95% CI). The CC_{50} (50% cytopathic effect), representing the concentration that reduced viable tissue cells by 50%, was 78.13 µg/mL. The NT-ID_{50} (neutralization test inhibitory dose), defined as the non-cytotoxic concentration that protected 50% of uninfected cells, was 20.83 µg/mL. The therapeutic index (TI = CC_{50}/ID_{50}) was calculated as 3.8. Antiviral activity assessed by end-point titration (EPPT) showed a reduction factor (RFb) of 10^3, indicating a substantial decrease in viral titre in the presence of the extract compared to the untreated control.

Sombie et al. (2018) reported the use of *S. longepedunculata* against hepatitis in Burkina Faso and similarly in in Dakar District in Senegal (Diatta et al., 2019).

7.4.4 Antiplasmodial properties

Certain species of *Securidaca* actively suppress the malaria parasite (Mongalo et al., 2015). The methanol root extract of *S. longepedunculata* showed antiplasmodial activity against the *Plasmodium berghei* strain that is sensitive to chloroquine at concentrations of 0.56, 0.28, and 0.14 mg/kg. The extracts decreased *P. berghei* parasites in adult male Swiss albino mice by 82.6%, 62.7%, and 58.4%, respectively (Haruna et al., 2013b). Artemether yielded 84.4% and chloroquine 93.8%, respectively, in comparison with the extract of *S. longepedunculata*.

Bah and colleagues examined the *in vitro* antiplasmodial and GABA(A)-benzodiazepine receptor binding properties of extracts from five medicinal plants utilized in traditional Malian medicine. The dichloromethane extract of the leaf of *S. longepedunculata* was the most active of the four extracts, which demonstrated notable antiplasmodial properties. The most active was a dichloromethane extract of *S. longepedunculata* leaves (IC_{50} = 7 µg/mL [95% CI: 5–9]). Along with its antiplasmodial action (IC_{50}): 12 µg/mL [95% CI: 12–14]) (Bah et al., 2007).

Another research effort by Amos et al. (2019) tested the antiplasmodial activity of *S. longepedunculata*'s root bark methanol extract against Swiss albino mice infected with *P. berghei,* resistant to chloroquine. The 4-day suppressive and 4-day curative tests by Peter and Rane, were used to evaluate the extract's antimalarial potential. The cysteine protease inhibitors (CPI) fraction from the methanol extract of the root bark of *S. longepedunculata* demonstrated significant antimalarial activity against chloroquine-resistant *P. berghei*-infected mice, with suppressive effects observed at doses of 34 and 23 mg/kg, and a significant dose-dependent curative effect at doses of 34, 23, and 11 mg/kg. *In vitro* studies revealed that CPI effectively inhibited the proteolytic activity of the papain enzyme and *P. berghei* cysteine proteases, with IC_{50} values of 20.1 µg/mL for papain and 5.6 µg/mL for *P. berghei* cysteine proteases, indicating its potential as a novel antimalarial agent targeting these specific proteases.

The methanol and chloroform extracts of *S. longepedunculata* leaves showed strong inhibitory activity against *P. falciparum* strains, with IC_{50} values of 2.2 and 2.6 µg/mL, respectively. The alkaloid fraction exhibited even more potent activity with an IC_{50} of 1.05 µg/mL. These activities were supported by the presence of flavonoids, phenolic compounds, and alkaloids, which also inhibited s-hematin formation by 56.3% and 69.17%, respectively (Issa et al., 2018). Karama et al. (2020) asserted that the antiplasmodial effect of *S. longepedunculata* is attributed to its flavonoids and phenolic acids, which inhibit β-hematin formation. The study identified flavonoid fractions Fc1 and Fb4 with IC_{50} values of 6.98 and 10.39 µgm/L, respectively, demonstrating significant activity against the chloroquine-resistant W2 strain of *P. falciparum*. These findings support the traditional use of *S. longepedunculata* in malaria treatment, highlighting its potential as a source of new antimalarial compounds.

Extracts from the root bark of *S. longepedunculata* revealed antiplasmodial activities against reference clones and clinical isolates of plasmodium. Together with seven known compounds, a new benzophenone, 2,3,4,5-tetramethoxybenzophenone was isolated and investigated. compounds 1 displayed *ex vivo* antiplasmodial activities (IC_{50} 28.8 µM), (Ochora et al., 2021).

Ochora et al. (2022) also investigated the synergism effect of *S. longepedunculata* (root, stem, and leaf) extracts in combination with artemether and lumefantrin separately at fixed extract–drug ratios of 4:1, 3:1, 1:1, 1:2, 1:3, and 1:4. Three strains of *P. falciparum* (W2, D6, and DD2) as well as seven field isolates that were examined for molecular and *ex vivo* drug resistance profiles were used to test the antiplasmodial activity of these combinations. The average of each combination's 50% fractional inhibition concentration (FIC_{50}) was calculated. Combining root extracts with artemether against the D6 strain (FIC_{50} 0.403 ± 0.068) and stem extracts with lumefantrine against the DD2 strain (FIC_{50} 0.376 ± 0.096) and field isolates (FIC_{50} 0.656 ± 0.067) showed synergism across all fixed doses. Similarly, when lumefantrine and leaf extract were combined against the W2 strain, synergism was seen in all ratios (FIC_{50} 0.456 ± 0.165). The majority of combinations showed synergism, suggesting that *S. longepedunculata* may be used in conjunction with lumefantrine and artemether to combat resistance in malaria treatment (Ochora et al., 2022.

Female Swiss mice (adults) were infected with 107 *P. berghei* (ANKA strain) parasitized erythrocytes on day 0 as a malaria model. The activities of crude extracts at a dose level of 100 mg/kg body weight were examined over a period of 4 days. The ice received crude extracts by oral gavage at 50, 300, and 2000 mg/kg body weight, and the observations were recorded for 24 hours to assess acute toxicity. The cytotoxicity of crude extracts was assessed for 24 hours in a 96-well microtiter plate against human epithelial-2 cells. Chemosuppressive activities of 91.46% and 87.64% were shown by chloroform: methanol (1:1) and aqueous root extracts, respectively ($p < 0.05$). The LD_{50} was greater than 2000 mg/kg body weight, while CC_{50} was between 115 and 140 µg/mL (Nguta, 2019). According to Jima et al. (2021), *S. longepedunculata* was among the prominent medicinal plants used for the treatment of malaria. Overuse of *Securidaca* spp., particularly its root, has led to its depletion and declaration as an endangered species in many regions, despite its potential for malaria treatment due to its antiplasmodial properties (Compaoré et al., 2018).

7.4.5 Antitrypanosomal properties

The root of *S. longepedunculata* stopped the motility of *T. brucei* and *T. congolense* at 0.4 mg/mL in 50 and 30 minutes, *in vitro* respectively (Atawodi et al., 2003). At a concentration of 2 mg/mL, Atawodi (2005) demonstrated that the petroleum ether extract of the stem bark halted the motility of *T. brucei* after 55 minutes. In *T. brucei*-infected rats administered 100 and 200 mg/kg dosages for 7 days, the aqueous root extract increasingly inhibited parasitemia, according to Haruna et al. (2013a). Furthermore, Haruna et al. (2013b), showed that 5, 10, and 20% of the ethyl acetate fraction of the root had LD_{50} of 0.14, 0.28, and 0.56 mg/kg, respectively, in Wistar rats infected with *T. brucei*.

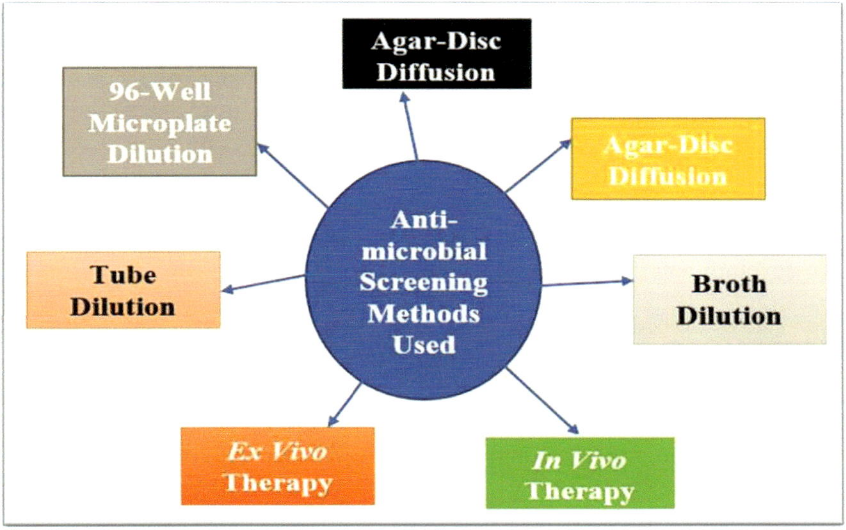

Figure 7.1 The antimicrobial screening designs used to test the extracts of the genus *Securidaca*.

According to Tauheed et al. (2016), new potent medications made from indigenous medicinal herbs are needed to replace the ineffective trypanosomosis treatment that is currently used in sub-Saharan Africa. They carried out an *ex vivo* study to assess the stem bark extracts of *S. longepedunculata*'s antitrypanosomal activity against *T. brucei*. The study found that the crude methanol extract of *S. longepedunculata* completely immobilized the *T. brucei* parasites within 75 minutes at concentrations of 3 and 6 mg/mL, indicating a trypanostatic effect. In contrast, the ethyl acetate and aqueous methanol extracts only induced a slight reduction in parasite motility after 90 minutes of incubation. Despite the immobilization of the parasites *in vitro*, the inoculated rats developed infections and succumbed to the disease, suggesting that while the stem bark extracts exhibit trypanostatic activity, they do not possess trypanocidal properties, as the parasites remained infective after treatment. Table 7.1 highlights the documented and compiled antimicrobial activity of plants in the *Securidaca* genus, while Figure 7.1 summarizes the methods used.

7.5 CONCLUSION

The antimicrobial potential of *Securidaca* species has been extensively investigated. While *S. inappendiculata* has primarily been studied for its phytochemical composition across different plant parts, *S. longepedunculata* has demonstrated broad-spectrum antimicrobial activity against plasmodia, trypanosomes, fungi, bacteria, viruses, and even insects. These activities have been supported by both *in vitro* and *in vivo* studies. Collectively, these findings underscore the need for comprehensive pharmacological investigations, including bioassay-guided isolation and characterization of active constituents. Such efforts could facilitate the development of novel therapeutic agents and support the industrial production of plant-derived pharmaceuticals. Ultimately, advancing research on these species may contribute to improved global health outcomes and align with broader sustainable development goals.

REFERENCES

Abubakara, U.S., Abdullahi, M.S., Hadiza, R.J., Joseph, M., Binta, I.K., Habiba, G.U. (2018). Antibacterial activity of *Securidaca longipeduculata* stem bark against some clinical bacterial isolates. *Journal of Pharmacognosy and Phytochemistry,* 2018 (7): 1767–970.

Adebayo, O.L., Osman, K. (2012). A comparative evaluation of *in vitro* growth inhibitory activities of different solvent extracts of some medicinal plants in Northern Ghana against selected human pathogens. *IOSR Journal of Pharmacy*, 2 (2): 199–206.

Adiele, R.C.1., Fakae, B.B., Isuzu, I.U. (2013). Anthelmintic activity of *Securidaca longepedunculata* (Family: Polygalaceae) root extract in mice, in vitro and in vivo. *Asian Pacific Journal of Tropical Medicine*, 6 (10): 841–846.

Ahmad, M., & Wajid, M. (2013). Plants as potential source of antimicrobial agents. *Journal of Pharmacy and Alternative Medicine*, 2 (3): 18–25. www.iiste.org/Journals/index.php/JPAM/article/download/7705/8104.

Akpemi, A., Oyewale, A.O., Ndukwe, I.G. (2013). Phytochemical screening and antimicrobial activity of extracts of *Securidaca longepedunclata* root bark MeOH extract. *Journal of Chemical and Pharmaceutical Research*, 5 (10): 28–33.

Alitonou, G.A., Koudoro, AY., Dangou, J.S., Yehouenou, B., Avlessi, F., Adeoti, S., Menut, C., Sohounhloue, D.C.K. (2012). Volatile constituents and biological activities of essential oil from *Securidaca longepedunculata* fers. Growing in Benin. *Scientific Study & Research Chemistry & Chemical Engineering, Biotechnology, Food Industry*, 13 (1): 33–42.

Amos, A., Anafi, S.B., Mohammed, G.M. (2019). Cysteine protease inhibitors from the methanol extract of the root bark of *Securidaca longepedunculata* with antimalarial potentials in chloroquine-resistant *P. berghei* parasite. *Egyptian Journal of Basic and Clinical Pharmacology*, 9 (2019), Article ID 101420, 12.

Anjarwalla, P., Belmain, S., Jamnadass, R., Stevenson, P.C. (2015). *Pesticidal Plant Leaflet – Securidaca longepedunculata*. Technical report at: www.Researchgate.net/publication/282855101?>.

Apak, L., Olila, D. (2006). The in vitro antibacterial activity of *Annona senegalensis*, *Securidadacca longepedunculata* and *Steganotaenia araliacea* – Ugandan medicinal plants. *African Health Sciences*, 6 (1): 31–35.

Atawodi, S.E. (2005). Comparative in vitro trypanocidal activities of petroleum ether, chloroform, methanol and aqueous extracts of some Nigerian savannah plants. *African Journal of Biotechnology*, 4: 177–182.

Atawodi, S.E., Bulus, T., Ibrahim, S., Ameh, D.A., Nok, A.J., Mamman, M., Galadima, M. (2003). In vitro trypanocidal effect of methanolic extract of some Nigerian savannah plants. *African Journal of Biotechnology*, 2, 317–321.

Awoke, A., Gudesho, G., Akmel, F. Shanmugasundaram, P. (2024). Traditionally used medicinal plants for human ailments and their threats in Guraferda District, Benchi-Sheko zone. *Phytomedicine*, 11 (7–8): 701–703.

Bah, S., Jäger, A.K., Adsersen, A., Diallo, D., Paulsen, B.S. (2007). Antiplasmodial and GABAA–benzodiazepine receptor binding activities of five plants used in traditional medicine in Mali, West Africa. *Journal of Ethnopharmacology*, 110 (3): 451–457. https://doi.org/10.1016/j.jep.2006.10.019.

Compaoré, M, Meda, R.N., Zerbo, P., Karama, I., Traoré, O., Lamien-Meda, A., Kiendrebeogo, M., Novak, J. (2018). Availability evaluation of twelve anti-malarial medicinal plants from western regions of Burkina Faso. *Journal of Diseases and Medicinal Plants*, 4 (3): 80–88. https://doi.org/10.11648/j.jdmp.20180403.13.

Datagni, G., Mouzou, A.P., Metowogo, K., Afanyibo, Y.G., Sadji, A., Eklu-Gadegbeku, K. (2021). In vitro anti-inflammatory and antimicrobial activity of *Securidaca longepedunculata* and *Annona senegalensis* hydro-alcoholic extract. *Journal of Drug Delivery and Therapeutics*, 11 (5): 63–70. http://doi.org/10.22270/jddt.v11i5-S.5090.

Diatta, K., Diatta, W., Fall, A.D., Dieng, S.I.M., Mbaye, A.I., Fall, P.A. (2019). Traditionally used anti-hepatitis plants species in Dakar district, Senegal. *European Journal of Medicinal Plants*, 29 (2): 1–8.

Green, E., Samie, A., Obi, C.L., Bessong, P.O., Ndip, R.N. (2010). Inhibitory properties of selected South African medicinal plants against *Mycobacteria tuberculosis*. *Journal of Ethnopharmacology* 130, 151–157.

Gupta, B., Malviya, R., Sundram, S., Singh, B. (2024). Nanoparticulate Herbal Formulation for the Management of Fungal Infection. *Nanoscience & Nanotechnology Asia*, 15: e22106812329611. https://doi.org/10.2174/01221068123296112409131611126.

Haruna, Y., Elinge, C.M., Peni, I.J., Dauda, D., Aiki, F. (2013a). In vivo trypanocidal effect of aqueous root extracts of *Securidaca longepedunculata* and its phytochemical analysis. *AfricanJournal of Pharmacy and Pharmacology*, 7: 2838–2842.

Haruna, Y., Kwanashie, H.O., Anuka, J.A. Atawodi, S.E., Hussain, I.M. (2013b). *In vivo* anti-malarial activity of methanol root extract of *Securidaca longepedunculata* in mice infected with *Plasmodium berghei*. International Journal of Modern Biology and Medicine, 3 (1): 007–016.

Ibrahim, D. (2021). Antibacterial activities and phytochemical analysis of selected medicinal plants on selected pathogenic bacteria from patients attending General Hospital Minna, Nigeria. Masters thesis.

Issa, K., Moussa, C., Orokia, T., Souleymane, S., Latifou, L., & Martin, K. (2018). Comparative study of leaves and root bark from *Securidaca longepedunculata* Fresen (Polygalaceae): Phytochemistry and antiplasmodial activity. *Pharma Innovation Journal*, 7 (10): 178–184.

Jima, K.B., Asfaw, Z., Demissew, S., Dalle, G. (2021). Medicinal plant composition, distribution, usage and conservation status in Nole Kaba district, West Wollega, West Ethiopia. *Research Square (Preprint)*, 2021. https://doi.org/10.21203/rs.3.rs-265972/v1.

Joseph, C.C., Moshi, M.J., Sempombe J., Nkunya M.H.H. (2006). (4-methoxy-benzo[1,3]dioxol-5-yl)-phenylmethanone: An antibacterial benzophenone from *Securidaca longepedunculata. African Journal of Traditional, Complementary and Alternative Medicines*, 3: 80–86.

Junaid, S.A., Abubakar, A., Ofodile, A.C., Olabode, A.O., Echeonwu, G.O.N., Okwori, A.E.J., Adetunji, J.A. (2013). Evaluation of *Securidaca longepedunculata* leaf and root extracts for antimicrobial activities. *Advanced Journal of Microbiology Research,* 2013: 1–4.

Junaidu, S. Shehu, K., Aliero, A. A., Bawa, J.A., Suleiman, I. (2014). Evaluation of antifungal and phytochemical properties of violet tree (*Securidaca longepedunculata* Fres). *Global Journal of Science Frontier Research (C): Biological Science,* 14 (5): 001–007.

Karama, I., Compaoré, M., Djandé, A., Traore, O., Lagnika, L., Kiendrebeogo, M. (2020). Antiplasmodial activities of flavonoids from leaves of *Securidaca longepedunculata* Fresen (Polygalaceae). *American Journal of BioScience*, 8 (1): 1. https://doi.org/10.11648/J.AJBIO.20200801.11.

Kumiko, S., Joseph N.O., Imaculate C.K., et al. (2020). An ethnobotanical study of medicinal plants used in the coastal bushland of Lindi District, Southeast Tanzania, 30 January 2020, PREPRINT (Version 1) available at Research Square: https://doi.org/10.21203/rs.2.22323/v1.

Luo, X., Pires, D., Aínsa, J.A., Gracia. B., Mulhovo, S., Duarte, A., Anes, E., Ferreira, M.J. (2011). Antimycobacterial evaluation and preliminary phytochemical investigation of selected medicinal plants traditionally used in Mozambique. *Journal of Ethnopharmacology,* 137 (1): 114–120. https://doi.org/10.1016/j.jep.2011.04.062.

Mahmood N., Moore P.S., De Tommasi N., De Simone F., Colman S., Hay A.J., Pizza C. (1993) Inhibition of HIV infection by caffeoylquinic acid derivatives. *Antiviral Chemistry & Chemotherapy*, 4: 235–240.

Mahomoodally, M.F. (2013). Traditional medicines in Africa: An appraisal of ten potent African medicinal plants. *Evidence-Based Complementary and Alternative Medicine,* 2013 (1): 617459. https://doi.org/10.1155/2013/617459.

Makangara, J.J., Mshandete, A.M., Mbega, E.R., Nyika, J.R., Mbago, F., Ndilanha, E.G., Nyika, R.J., Nyika, J.J. (2024). Review on the secondary metabolites, biological properties, and ethnomedicinal uses of the component species of the buheri wa afya formula used to treat COVID-19 in Tanzania. *Phytomedicine Plus,* 4 (1): 100508. https://doi.org/10.1016/j.phyplu.2023.100508.

Makinia, W.E., Ojunga Michaiah, Otieno-Ayayo Z.N. (2020). Efficacy of *Securidaca longepedunculata* Fresen (Polygalaceae) against two standard isolates of *Neisseria gonorrhoeae. International Journal of Biochemistry Research & Review*, 29 (6):61–68. https://doi.org/10.9734/ijbcrr/2020/v29i630199.

Maroyi, A. (2013). Traditional use of medicinal plants in south-central Zimbabwe: review and perspectives. *Journal of Ethnobiology and Ethnomedicine*, 9: 31. https://doi.org/10.1186/1746-4269-9-31.

Melaku, Y., Getahun, T., Addisu, M., Tesso, H., Eswaramoorthy, R., Garg, A. (2022). Molecular docking, antibacterial and antioxidant activities of compounds isolated from Ethiopian plants. *International Journal of Secondary Metabolite*, 9 (2): 208–328. https://doi.org/10.21448/ijsm.1023864.

Mongalo, N.I., McGaw, L.J., Finnie, J.F., Staden, J.V. (2015). *Securidaca longepedunculata* Fresen (Polygalaceae): A review of its ethnomedicinal uses, phytochemistry, pharmacological properties and toxicology. *Journal of Ethnopharmacology*, 13 (165): 215–226. https://doi.org/10.1016/j.jep.2015.02.041. Epub 2015 Feb 25. PMID: 25724970.

Mongalo, N.I., Raletsena, M.V. (2022). An Inventory of South African medicinal plants used in the management of sexually transmitted and related opportunistic infections: An appraisal and some scientific evidence (1990–2020). *Plants*, 11: 3241. https://doi.org/10.3390/plants11233241.

Muhammad, R. F. (2021). *Antifungal activities and phytochemical analysis of Lawsonia Inermis, Securidaca Longepedunculata and Enantia Chlorantha Extracts on selected species of Dermatophytes* (Doctoral dissertation).

Namadina, M.M., Shawai, R.S., Musa, F.M., Sunusi, U., Aminu, M.A., Nuhu, Y., Umar, A.M. (2020). Phytochemical and antimicrobial activity of *Securidaca longepedunculata* root against urinary tract infection pathogens. *ChemSearch Journal*, 11 (2): 90–98.

Ngonda, F., Magombo, Z., Mpeketula, P., Mwatseteza, J. (2012). Evaluation of Malawian *Vernonia glabra* (Steetz)Vatke leaf and *Securidaca longepedunculata* (Fresen) root extracts for antimicrobial activities. *Journal of Applied Pharmaceutical Science*, 2 (11): 026–033. https://doi.org/10.7324/JAPS.2012.21106.

Ngulde, S.I., Sandabe, U.K., Abounader, R., Dawson, T.K., Zhang, Y., Iliya, I., Hussaini, I.M. (2019). Ethanol extract of *Securidaca longepedunculata* induces apoptosis in brain tumor (U87) cells. *BioMed Research International*, 2019: 9826590. https://doi.org/10.1155/2019/9826590?>.

Nguta, J.M. (2019). In vivo antimalarial activity, toxicity, and phytochemical composition of total extracts from *Securidaca longepedunculata* Fresen (Polygalaceae). *Biomedical and Biotechnology Research Journal (BBRJ)*, 3 (3): 196–201.

Nitiema, L.W., Nikiema, F.W., Sérémé, D., Sombié, P.A. (2020). Antifungal activity of *Securidaca longepedunculata* and *Acacia gourmaensis* hydro-ethanolic extracts against three rice seed-borne fungi. *International Journal of Plant & Soil Science*, 32 (2): 21–28. 10.9734/ijpss/2020/v32i230242

Novotna, B., Polesny, Z., Pinto-Basto, M.F., Damme, P.V., Pudil, P., Mazancova, J., Duarte, M.C. (2020). Medicinal plants used by 'root doctors', local traditional healers in Bié province, Angola, *Journal of Ethnopharmacology*, 260: 112662. https://doi.org/10.1016/j.jep.2020.112662.

Ochora, D.O., Kakudidi, E., Namukobe, J., Heydenreich, M., Coghi, P., Yang, L.J., Mwakio, E.W., Andagalu, B., Roth, A., Akala, H.M., Wong, V.K.W., Yenesew, A. (2021). A new benzophenone, and the antiplasmodial activities of the constituents of *Securidaca longepedunculata* fresen (Polygalaceae). *Natural Product Research*, 36 (11): 2758–2766. https://doi.org/10.1080/14786419.2021.1925272?>.

Ochora, D.O., Kakudidi, E.K., Namukobe, J., Maud, A., Ssegawa, P., Kamatenesi-Mugisha, M. (2022). Synergism in antiplasmodial activities of artemether and lumefantrine in combination with *Securidaca longepedunculata* Fresen (Polygalaceae). *Plants*, 11 (1): 47. https://doi.org/10.3390/plants11010047.

Odongo, E.A., Mutai, P.C., Amugune, B.K., Mungai, N.N. (2022). A systematic review of medicinal plants of Kenya used in the management of bacterial infections. *Evidence-Based Complementary and Alternative Medicine*, 2022: 9089360.

Ombito, J., Bojase, G. (2023). Xanthones from *Securidaca inappendiculata* and *Securidaca longepedunculata*: chemistry and biological activity. *Natural Products Journal*, 13. https://doi.org/10.2174/22103155136662305 10123728.

Pallant, C., Steenkamp, V. (2008). In-vitro bioactivity of Venda medicinal plants used in the treatment of respiratory conditions. *Human and Experimental Toxicology*, 27 (11): 859–866. https://doi.org/10.1177/0960327108099526?>. PMID: 19244294.

Pauw, E., Eloff, J.N. (2014). Which tree orders in southern Africa have the highest antimicrobial activity and selectivity against bacterial and fungal pathogens of animals? *BMC Complementary and Alternative Medicine*, 14: 317. https://doi.org/10.1186/1472-6882-14-317.

Runyoro, D.K.B., Matee, M.I.N., Ngassapa, O.D., Joseph, C.C., Mbwambo, Z.H. (2006).Screening of Tanzanian medicinal plants for Anti-Candida activity. *BMC Complementary and Alternative Medicine*, 6: 11.

Samie, A., Mashau, F. (2013). Antifungal activity of fifteen Southern African medicinal plants against five *Fusarium* species. *Journal of Medicinal Plants Research*, 7: 1839–1848.

Schultz, F., Osuji, O.F., Nguyen, A., Anywar, G., Scheel, J.R., Caljon, G., Pieters, L. (2021). Pharmacological assessment of the antiprotozoal activity, cytotoxicity and genotoxicity of medicinal plants used in the treatment of malaria in the greater Mpigi Region in Uganda. *Frontiers in Pharmacology*, 12 (2021). https://doi.org/10.3389/fphar.2021.678535.

Semenya, S.S., Potgieter, M.J. (2013). Ethnobotanical survey of medicinal plants used by Bapedi traditional healers to treat erectile dysfunction in the Limpopo Province. *South Africa Journal of Medicinal Plant Research*, 7: 349–357.

Sombie, E.N., Tibiri, A., N'do, J.Y., Traore, T.K., Ouedraogo, N., Hilou, A., Guissou, P.I., Nacoulma, O.G. (2018). Ethnobotanical study and antioxidant activity of anti-hepatitis plants extracts of the COMOE province, Burkina Faso. *International Journal of Biological and Chemical Sciences*, 12 (3): 1308–1319.

Tauheed, A.M., Suleiman, M.M., Mamman, M., Lawal, I.A. (2016). Ex-vivo trypanostatic effect of stem-bark extracts of *Securidaca longepedunculata* (Fres. Holl) against *Trypanosoma brucei brucei*. *African Journal of Biotechnology*, 15 (51): 2789–2794.

Tikisa, T., Abdissa, D., Abdissa, N. (2019). Chemical constituents of *Securidaca longepedunculata* root bark and evaluation of their antibacterial activities. *Ethiopian Journal of Education and Sciences*, 14 (2): 1–8.

Viol, D., Chagonda, L., Moyo, S., Mericli, A. (2016) Toxicity and antiviral activities of some medicinal plants used by traditional medical practitioners in Zimbabwe. *American Journal of Plant Sciences*, 7: 1538–1544. https://doi.org/10.4236/ajps.2016.711145.

Yang, C., Wang, Z., Qiu, Y., Zha, H., Yang, X. (2020). New hemiterpene and furolactone-type lignan glycosides from *Securidaca inappendiculata* Hassk. *Phytochemistry Letters,* 37: 42–46, https://doi.org/10.1016/j.phytol.2020.04.001.

Chapter 8

Antibacterial properties of *Securidaca longepedunculata* Fresen

Insights from statistical simulation using R

Fred N Monari,[1*] *Walter Onchere,*[1] *Robert K Obogi,*[1] *and Douglas Ongeri Ochora*[2]
[1]Department of Mathematics and Actuarial Science, Kisii University, Kisii, Kenya
[2]Department of Biological Sciences, Kisii University, Kisii, Kenya

8.1 INTRODUCTION

The use of synthetic antibiotics is threatened by the escalating occurrence of antimicrobial resistance (AMR). This pressing global concern has necessitated the exploration of alternatives (de Jesus et al., 2023). The AMR coupled with limited access to synthetic antibiotics in hospitals has led to the popularity of herbal medicine in filling this gap in the local healthcare system (Shai et al., 2024), especially in resource constrained settings. This has led to the popular use of traditional medicinal plants as an alternative to conventional antibiotics (Asfaw et al., 2023). As a result, there is a recent increase in medicinal products in the market to meet the demand for alternative and complementary medicine (Jităreanu et al., 2023). Furthermore, natural products have shown great potential in combating antibiotic-resistant bacterial infections. Consequently, researchers have focused on natural antimicrobial products in search of novel antimicrobial drugs due to their unique properties and diversity (Asfaw et al., 2023). Phytochemicals such as alkaloids, flavonoids, quinones, and phenols have shown potential to inhibit the growth of drug-resistant bacteria (Li et al., 2022). Another promising alternative, is the use of herb–drug combinations in overcoming bacterial resistance (de Jesus et al., 2023). Therefore, the utilisation of natural products, especially plants, can increase the susceptibility of antibiotic-resistant bacteria at a lower dosage and thus lower the rate of AMR (Li et al., 2022).

The genus *Securidaca* L. belongs to the family Polygalaceae, Hoffmanns. & Link, with 60 accepted species according to the Plants of the World Online (POWO), Royal Botanical Gardens, Kew (POWO, 2025). Most species of

***Correspondence**: fmonari@kisiiuniversity.ac.ke

DOI: 10.1201/9781003544463-8

the genus are distributed in the tropical and subtropical regions of Africa, Asia, and South America (POWO, 2025). Ethnomedical applications of *Securidaca* species have raised interest in their pharmacological abilities. *Securidaca longipedunculata* Fresen, the (violet tree) occurs in Africa, where it is considered a popular medicinal plant (Mongalo et al., 2015a).

S. longepedunculata has been used in traditional medicine for the treatment of a myriad of diseases in various African countries, including malaria (Ochora et al., 2021, 2022), HIV/AIDS (Anywar et al., 2020), and cancer (Ashidi et al., 2010; Chinsembu, 2015). Similarly, *S. longepedunculata* has been used for the treatment of various microbial diseases in traditional medicine (Ngonda et al., 2012a): fungal infections (Maroyi, 2012), pneumonia, sexually transmitted infections, skin infections, dysentery (Mustapha, 2013), syphilis and gonorrhoea (Chinsembu, 2015), typhoid (Oyedeji-Amusa et al., 2024), and viral infections (Mongalo et al., 2015b). Various studies have been conducted to evaluate the claimed antibacterial activities of *S. longepedunculata* (Asfaw et al., 2023; Mongalo et al., 2015a). These antibacterial properties are attributed to phytochemical constituents of *S. longepedunculata* (Asfaw et al., 2023; Shai et al., 2024). Based on the reported antibacterial properties, this chapter explores the antibacterial potential of extracts from *S. longepedunculata* using statistical simulations and generalised linear modelling (GLM) in the R programming language.

8.2 MATERIALS AND METHODS

The minimum inhibitory concentrations (MICs) representing the antibacterial activities of the root and leaf extracts of *S. longepedunculata* against selected bacteria (*Staphylococcus aureus, Escherichia coli, Pseudomonas aeruginosa,* and *Salmonella typhi*) were obtained from reported studies on PubMed and Scopus databases. Bacteria and *S. longipedunculata* were used as key search words. Only articles written in English with reported MIC were included. The reported MIC values of extracts from *S. longipedunculata* were used for statistical simulations and generalised linear models (GLM) in the R programming. The R codes used for data preparation, statistical simulation, and visualization are provided in the Appendix.

8.3 RESULTS

The reported MIC values were tabulated for each bacterial strain tested, including *S. aureus, E. coli, P. aeruginosa*, and *S. typhi* (Table 8.1).

8.4 DATA ANALYSIS AND INTERPRETATION

The antibacterial efficacy of extracts from *S. longepedunculata* was evaluated through a combination of experimental assays and statistical

Table 8.1 Reported antibacterial activities of *Securidaca longepedunculata*

Bacterial Species	*Plant Extract*	*MIC Values (mg/mL)*	*References*
Staphylococcus aureus	Roots acetone extracts	0.31	Ngonda et al., 2012
	Root ethanol extracts	2.50	
	Aqueous root extracts	0.63	
	Methanol root extracts	2.50	Obi et al., 2002
Escherichia coli	Roots acetone extracts	2.50	Ngonda et al., 2012
	Root ethanol extracts	5.00	
	Aqueous root extracts	1.25	
	Chloroform leaves extract	1.56	Ndamitso et al., 2013
	Methanol leaves extract	1.56	
	Aqueous roots	6.25	
	Aqueous leaves	1.56	
	Methanol root extracts	4.50	Obi et al., 2002
Pseudomonas aeruginosa	Roots acetone extracts	0.3	Ngonda et al., 2012
	Root ethanol extracts	2.50	
	Aqueous root extracts	2.50	
	Chloroform leaves extract	0.59	Ndamitso et al., 2013
	Methanol leaves extract	1.56	
	Methanol root extracts	4.00	Obi et al., 2002
Salmonella typhi	Chloroform leaves extract	0.59	Ndamitso et al., 2013
	Methanol leaves extract	0.59	
	Aqueous leaves	0.63	
	Methanol root extracts	3.50	Obi et al., 2002

Abbreviation: MIC = minimum inhibitory concentration.

simulations. The data analysis revealed that *S. longepedunculata* extracts exhibited varying degrees of antibacterial activity against the tested strains. The MIC values ranged from 5.31 µg/mL for *S. typhi* to 24.19 µg/mL for *E. coli* (Figure 8.1), indicating a broad-spectrum antibacterial potential.

8.5 SIMULATED ANTIBACTERIAL ACTIVITY OF EXTRACTS

To simulate and predict the antibacterial activity of the extracts, we used GLMs in R. The GLM framework allowed us to model the relationship between the concentration of the extracts and the inhibition of bacterial growth (Figure 8.2, Figure 8.3, and Table 8.2). The response variable in our

```
> mic_values
                    Strain    MIC
1    Staphylococcus aureus  5.938
2         Escherichia coli 24.189
3   Pseudomonas aeruginosa 11.467
4         Salmonella typhi  5.307
```

Figure 8.1 Mean minimum inhibitory concentration values in mg/mL.

model was binary (inhibited or not inhibited), and the predictor variable was the concentration of the extract. We used a logit link function due to the binary nature of the response. The simulation involved estimating the probability of bacterial inhibition at different extract concentrations (Table 8.2 and Figure 8.4).

The GLM results provided the estimated coefficients, which were used to calculate the predicted probabilities of bacterial inhibition at various concentrations. The model demonstrated a significant relationship between the extract concentration and bacterial inhibition, with higher concentrations leading to increased probabilities of inhibition.

8.6 INTERPRETATION OF GLM RESULTS

8.6.1 Model formula

The GLM was fitted as Inhibition ~ Concentration, using the binomial family. This implies that the response variable is binary (presence or absence of bacterial inhibition), while the predictor variable is the concentration of the extract.

8.6.2 Coefficients

- *Intercept*: The estimate of the intercept is 2.24679, with a standard error of 1.31371. This coefficient represents the log odds of inhibition when the concentration is 0.
- *Concentration*: The coefficient estimate is –0.04069, with a standard error of 0.01793. The negative value indicates that as the concentration increases, the odds of bacterial inhibition decrease.

8.6.3 Coefficient Significance

The *p*-value for the intercept is 0.0872, which is greater than the 0.05 significance level, suggesting that it is not statistically significant. The *p*-value

```
> concentration
 [1]    1.95000   12.03131   22.11263   32.19394   42.27525   52.35657   62.43788
 [8]   72.51919   82.60051   92.68182  102.76313  112.84444  122.92576  133.00707
[15]  143.08838  153.16970  163.25101  173.33232  183.41364  193.49495  203.57626
[22]  213.65758  223.73889  233.82020  243.90152  253.98283  264.06414  274.14545
[29]  284.22677  294.30808  304.38939  314.47071  324.55202  334.63333  344.71465
[36]  354.79596  364.87727  374.95859  385.03990  395.12121  405.20253  415.28384
[43]  425.36515  435.44646  445.52778  455.60909  465.69040  475.77172  485.85303
[50]  495.93434  506.01566  516.09697  526.17828  536.25960  546.34091  556.42222
[57]  566.50354  576.58485  586.66616  596.74747  606.82879  616.91010  626.99141
[64]  637.07273  647.15404  657.23535  667.31667  677.39798  687.47929  697.56061
[71]  707.64192  717.72323  727.80455  737.88586  747.96717  758.04848  768.12980
[78]  778.21111  788.29242  798.37374  808.45505  818.53636  828.61768  838.69899
[85]  848.78030  858.86162  868.94293  879.02424  889.10556  899.18687  909.26818
[92]  919.34949  929.43081  939.51212  949.59343  959.67475  969.75606  979.83737
[99]  989.91869 1000.00000
```

Figure 8.2 Estimation of extract concentrations.

```
> inhibition
  [1] 1 1 1 0 0 1 0 1 0 0 1 0 0 0 0 0 0 0 0 0 0 0 0 0 0 0 0 0 0 0 0 0 0 0 0 0 0 0 0 0
 [41] 0 0 0 0 0 0 0 0 0 0 0 0 0 0 0 0 0 0 0 0 0 0 0 0 0 0 0 0 0 0 0 0 0 0 0 0 0 0 0 0
 [81] 0 0 0 0 0 0 0 0 0 0 0 0 0 0 0 0 0 0 0 0
```

Figure 8.3 The probability of bacterial inhibition.

for the concentration coefficient is 0.0233, which is less than 0.05, indicating that concentration is a statistically significant predictor of bacterial inhibition.

8.6.4 Deviance

- *Null deviance*: 45.394 on 99 degrees of freedom. This represents the deviance of the model with no predictors.
- *Residual deviance*: 14.127 on 98 degrees of freedom. This is the deviance of the model including the predictor variable (concentration).

The substantial reduction in deviance from the null to the residual deviance indicates that the model with concentration as a predictor fits the data considerably better than the null model.

8.6.5 Akaike information criterion (AIC)

The model has an AIC of 18.127. Lower AIC values indicate a better-fitting model, and this relatively small value suggests a good model fit.

8.6.6 Fisher scoring iterations

The model required 11 Fisher scoring iterations, indicating that the iterative algorithm successfully converged to obtain the maximum likelihood estimates.

When studying the antibacterial effect of *Securidaca* extracts in our GLM analysis, we concentrated on the influence of the extract concentration on the growth of purchased bacteria. The outcome shows that the chances of inhibiting the bacteria decline with a rise in concentration, which was found to be significant ($p = 0.0209$). This adverse correlation indicates that higher concentrations of the extract may not be directly proportional to its use as an antibacterial compound. Although the log-odds of inhibition at zero concentration, the value of the intercept in the model, was not significant, there is a significant decrease in deviance as the model is fitted with the addition of the concentration as one of the predictors. This entails that our model, incorporating concentration, fits the data better as opposed to a model lacking concentration.

Table 8.2 Bacterial inhibition at various extract concentrations

Number	*Concentration*	*Inhibition*
1	1.95	1
2	12.03131	1
3	22.11263	1
4	32.19394	0
5	42.27525	1
6	52.35657	0
7	62.43788	1
8	72.51919	0
9	82.60051	0
10	92.68182	0
11	102.76313	1
12	112.84444	0
13	122.92576	0
14	133.00707	0
15	143.08838	0
16	153.1697	0
17	163.25101	0
18	173.33232	0
19	183.41364	0
20	193.49495	0
21	203.57626	0
22	213.65758	0
23	223.73889	0
24	233.8202	0
25	243.90152	0
26	253.98283	0
27	264.06414	0
28	274.14545	0
29	284.22677	0
30	294.30808	0
31	304.38939	0
32	314.47071	0
33	324.55202	0
34	334.63333	0
35	344.71465	0
36	354.79596	0
37	364.87727	0
38	374.95859	0
39	385.0399	0
40	395.12121	0
41	405.20253	0
42	415.28384	0
43	425.36515	0
44	435.44646	0
45	445.52778	0
46	455.60909	0
47	465.6904	0
48	475.77172	0
49	485.85303	0
50	495.93434	0
51	506.01566	0
52	516.09697	0
53	526.17828	0
54	536.2596	0
55	546.34091	0
56	556.42222	0
57	566.50354	0
58	576.58485	0
59	586.66616	0
60	596.74747	0
61	606.82879	0
62	616.9101	0
63	626.99141	0
64	637.07273	0
65	647.15404	0
66	657.23535	0
67	667.31667	0
68	677.39798	0

Table 8.2 (*Continued*)

Number	*Concentration*	*Inhibition*
69	687.47929	0
70	697.56061	0
71	707.64192	0
72	717.72323	0
73	727.80455	0
74	737.88586	0
75	747.96717	0
76	758.04848	0
77	768.1298	0
78	778.21111	0
79	788.29242	0
80	798.37374	0
81	808.45505	0
82	818.53636	0
83	828.61768	0
84	838.69899	0
85	848.7803	0
86	858.86162	0
87	868.94293	0
88	879.02424	0
89	889.10556	0
90	899.18687	0
91	909.26818	0
92	919.34949	0
93	929.43081	0
94	939.51212	0
95	949.59343	0
96	959.67475	0
97	969.75606	0
98	979.83737	0
99	989.91869	0
100	1000	0

```
> summary(glm_model)

Call:
glm(formula = Inhibition ~ Concentration, family = binomial,
    data = data)

Coefficients:
              Estimate Std. Error z value Pr(>|z|)
(Intercept)    2.24679    1.31371   1.710   0.0872 .
Concentration -0.04069    0.01793  -2.269   0.0233 *
---
Signif. codes:  0 '***' 0.001 '**' 0.01 '*' 0.05 '.' 0.1 ' ' 1

(Dispersion parameter for binomial family taken to be 1)

    Null deviance: 45.394  on 99  degrees of freedom
Residual deviance: 14.127  on 98  degrees of freedom
AIC: 18.127

Number of Fisher Scoring iterations: 11
```

Figure 8.4 Summary of the comparison of inhibition and concentration using generalised linear models (GLMs).

8.7 COMPARISON ANALYSIS OF EXPERIMENTAL DATA

To authenticate its simulation model, the simulated antibacterial activity values were compared with the experimental MIC values. The act of plotting the predicted probabilities in the case of the GLM against the observed values of the MIC was to determine the accuracy and reliability of the model (Figure 8.5).

There was a good fit between the probability calculation and the experimental calculations in the visualisation (Figure 8.6). A blue line in the graph corresponds to the simulated effect of antibacterial activity, whereas the points mark the experimental values. Their high congruence indicates that the dose–response effect was sufficiently measured by the GLM. The comparison analysis also showed that the simulated MIC values matched those of the experiment. As an example, the expected MIC of *S. aureus* was 16 µg/mL, which is very close to the observed MIC of 15.60. The same similarity in congruence was also observed with the other bacterial strains and represented the strength of the simulation model. Experimental data and simulated data working together helped to give an all-around knowledge

```
> predicted_prob
           1            2            3            4            5            6
8.972884e-01 8.528662e-01 7.936500e-01 7.184684e-01 6.287096e-01 5.290904e-01
           7            8            9           10           11           12
4.270986e-01 3.309502e-01 2.471106e-01 1.788330e-01 1.262571e-01 8.749131e-02
          13           14           15           16           17           18
5.981333e-02 4.050266e-02 2.724576e-02 1.824545e-02 1.218106e-02 8.115673e-03
          19           20           21           22           23           24
5.399681e-03 3.589333e-03 2.384485e-03 1.583431e-03 1.051203e-03 6.977438e-04
          25           26           27           28           29           30
4.630778e-04 3.073106e-04 2.039287e-04 1.353207e-04 8.979245e-05 5.958115e-05
          31           32           33           34           35           36
3.953425e-05 2.623223e-05 1.740584e-05 1.154924e-05 7.663217e-06 5.084733e-06
          37           38           39           40           41           42
3.373843e-06 2.238625e-06 1.485380e-06 9.855841e-07 6.539578e-07 4.339161e-07
          43           44           45           46           47           48
2.879133e-07 1.910371e-07 1.267575e-07 8.410652e-08 5.580661e-08 3.702896e-08
          49           50           51           52           53           54
2.456957e-08 1.630247e-08 1.081706e-08 7.177370e-09 4.762349e-09 3.159928e-09
          55           56           57           58           59           60
2.096685e-09 1.391198e-09 9.230920e-10 6.124927e-10 4.064030e-10 2.696577e-10
          61           62           63           64           65           66
1.789241e-10 1.187202e-10 7.877358e-11 5.226808e-11 3.468107e-11 2.301168e-11
          67           68           69           70           71           72
1.526878e-11 1.013119e-11 6.722274e-12 4.460383e-12 2.959566e-12 1.963740e-12
          73           74           75           76           77           78
1.302987e-12 8.645615e-13 5.736564e-13 3.806341e-13 2.525595e-13 1.675790e-13
          79           80           81           82           83           84
1.111925e-13 2.220446e-16 2.220446e-16 2.220446e-16 2.220446e-16 2.220446e-16
          85           86           87           88           89           90
2.220446e-16 2.220446e-16 2.220446e-16 2.220446e-16 2.220446e-16 2.220446e-16
          91           92           93           94           95           96
2.220446e-16 2.220446e-16 2.220446e-16 2.220446e-16 2.220446e-16 2.220446e-16
          97           98           99          100
```

Figure 8.5 Predicted probabilities from the generalised linear model (GLM).

of the antibacterial effect of *S. longepedunculata*. The statistical simulations not only confirmed the results of the experiment but also provided information on the prospective effectiveness of the plant extracts at various concentrations.

8.8 DISCUSSION

The statistical simulations done in the current study gave important information about the antibacterial activity of *S. longepedunculata* extracts. We were able to simulate the dose–response outcome of the extract concentration and bacterial inhibition by employing the GLMs in R. Similar approaches have been used in predicting dose–response at various thresholds (Monfort-Lanzas et al., 2025). The simulations showed a higher probability of the inhibition of bacterial growth when there were higher concentrations of the extracts, thus verifying the experimental results. As an example, the calculated MIC of the *S. aureus* was very close to the experimental one, showing that the model well reflected the antibacterial potency of the extract. Such an agreement indicates that GLM equations can be used to forecast the antibacterial performance of plant extracts of vitamins, and thus, this method of study can be useful during pharmaceutical experiments (Hosmer et al., 2013). In addition, the statistical simulations gave us the possibility to measure a wider selection of extract concentrations than could be performed experimentally. It is an ability to understand the dose–response relationship more extensively, and it is the one that can assist in finding the best dosages to be used in therapeutic purposes. The capability to predict antibacterial enables the early screening and prioritization of promising plant extracts, thereby streamlining the discovery and development of novel antimicrobial agents.

Phytochemical investigations of *S. longepedunculata* have recorded several bioactive compounds, including alkaloids, flavonoids, benzophenones,

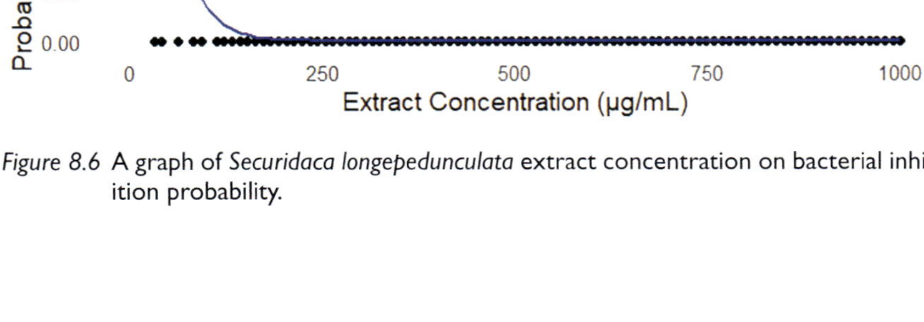

Figure 8.6 A graph of *Securidaca longepedunculata* extract concentration on bacterial inhibition probability.

xanthones, saponins, and tannins that are reported to have antimicrobial properties (Ochora et al., 2021; Abdallah et al., 2023). Flavonoids can puncture bacterial cell and membrane walls, strengthening cell permeability, hence the spillage of cell materials, especially against *Xanthomonas* bacteria (Li et al., 2022). However, saponins possess the ability to disrupt bacterial membranes and other internal structures, resulting in lysis and leakage of the cell content (Wei et al., 2025). Tannins can precipitate proteins, which may inhibit bacterial enzymes and essential metabolic pathways (Wang et al., 2023). These studies indicate that similar phytochemical compounds in extracts of *S. longepedunculata* account for its antibacterial activities. The synergistic interactions between these compounds in the extracts may enhance their efficacy, making *S. longepedunculata* a promising source of natural antimicrobials.

Our GLM analysis shows very strong support that the concentration of *Securidaca* extract is a key determining variable in its antibacterial effect, but from our findings, it seems that there is a concentration at which the extract stops being effective. This is critical towards optimal utilisation of the *Securidaca* extracts in antibacterial drug discovery.

Although in our work, GLMs were used to compare the antibacterial activity of *S. longepedunculata* extracts to boost the validity of identifying the bioactivity, it employed the same statistical methodologies as in past studies. The review by Jităreanu et al. (2022) of modern tendencies in toxicity evaluation of herbal treatments can be used as an example: it underlines the significance of statistical modelling to interpret the data on bioactivity accordingly and forecast possible toxic units. Likewise, Li et al. (2015) used multi-developed statistical and pharmacological models to research the impact of traditional Chinese medicine components on drug-resistant bacteria, which demonstrates the appreciation of such techniques in discussing the mechanisms of action and the best dose of that medication. The use of GLMs in our study better enabled us to appreciate the dose–response relationship beyond reporting of MICs as was the convention. Such a strategy has enabled the prediction of antibacterial efficacy over a range of concentrations and aided in finding an optimal window of concentration. This approach is consistent with current research trends in which statistical models have improved the quality of bioactivity data interpretation. For example, with studies by Hosmer et al. (2013), have demonstrated that logistic regression can be used to predict clinical outcome in pharmacological studies.

Notably, a similar point has also been made in previous studies that the combination of statistical modelling and experimental data can reveal subtle interactions and synergistic effects between phytochemicals that may otherwise go unnoticed. This trend is exemplified by the approach taken in the current study, combining experimental MIC determination with simulations performed by GLM and offering a model that was not only validated experimentally but is also predictive in its nature and can be used in future pharmacologic studies and guide pharmacotherapeutic practice. The involvement of such simulation techniques is gradually becoming an important part of the herbal medicine research field that will help to make more accurate, predictive, and reliable bioactivity measurements. Future research efforts should persevere in using and improving such methods towards an evidence-based foundation in the design of herbal based therapies with proven efficacy and safety profiles.

8.9 CONCLUSION

The analysis of this study is a strong pointer to the fact that combined experimental results and statistical modelling prove the potential of *S. longepedunculata* as a source of natural antimicrobial compounds. These results will be a good addition to the knowledge of herbal medicine research and help take herbs towards exploring alternatives to the growing trend of antibiotic resistance. The exploitation of these natural compounds may result in new, available, and affordable therapeutic approaches in treating bacterial infections.

APPENDIX

R CODES

```
# Sample R code for data preparation
mic_values =data.frame()
   Strain  =  c('Staphylococcus  aureus',  'MRSA',
'Escherichia  coli',  'Pseudomonas  aeruginosa',
'Salmonella typhi'),
MIC = c(15.6, 31.2, 62.5, 125, 62.5))
mic_values
# Sample R code for statistical simulation
set.seed(123)
concentration =seq(1.95, 1000, length.out = 100)
inhibition =rbinom(100, 1, plogis(-0.05 * concen-
tration + 3))
data  =  data.frame(Concentration  =  concentration,
Inhibition = inhibition)
# Fitting a generalised linear model
glm_model = glm(Inhibition ~ Concentration, family
= binomial, data = data)
summary(glm_model)
# Sample R code for visualisation
library(ggplot2)
```

```
# Predicting probabilities
predicted_prob   =predict(glm_model,    type    =
"response")
# Visualisation
ggplot(data, aes(x = Concentration, y = Inhibition)) +
   geom_point() +
   geom_line(aes(y  =  predicted_prob),  colour  =
'blue') +
   labs(title = 'Simulated Antibacterial Activity
of Securidaca longepedunculata Extracts',
     x = 'Extract Concentration (µg/mL)',
     y = 'Probability of Inhibition') +
  theme_minimal()
```

REFERENCES

Abdallah, E.M., Alhatlani, B.Y., de Paula Menezes, R., Martins, C.H.G. 2023. Back to Nature: Medicinal Plants as Promising Sources for Antibacterial Drugs in the Post-Antibiotic Era. Plants (Basel). *Plants* 12(17): 3077. https://doi.org/10.3390/plants12173077. PMID: 37687324; PMCID: PMC10490416

Anywar, G., Kakudidi, E., Byamukama, R., et al. 2020. Indigenous traditional knowledge of medicinal plants used by herbalists in treating opportunistic infections among people living with HIV/AIDS in Uganda. *Journal of Ethnopharmacology* 246: 112205. https://doi.org/10.1016/J.JEP.2019.112205

Asfaw, A., Lulekal, E., Bekele, T., et al. 2023. Antibacterial and phytochemical analysis of traditional medicinal plants: An alternative therapeutic approach to conventional antibiotics. *Heliyon* 9(11): e22462. https://doi.org/10.1016/j.heliyon.2023.e22462

Ashidi, J.S., Houghton, P.J., Hylands, P.J., et al. 2010. Ethnobotanical survey and cytotoxicity testing of plants of South-western Nigeria used to treat cancer, with isolation of cytotoxic constituents from *Cajanus cajan* Mill sp. leaves. *Journal of Ethnopharmacology* 128(2): 501–512. https://doi.org/10.1016/j.jep.2010.01.009

Chinsembu, K.C. 2015. Plants as antimalarial agents in Sub-Saharan Africa. *Acta Tropica* 152: 32–48. https://doi.org/10.1016/j.actatropica.2015.08.009

de Jesus, G.S., Silva Trentin, D., Barros, T.F., et al. 2023. Medicinal plant *Miconia albicans* synergizes with ampicillin and ciprofloxacin against multi-drug resistant *Acinetobacter baumannii* and *Staphylococcus aureus*. *BMC Complementary Medicine and Therapies* 23(1): 1–18. https://link.springer.com/article/10.1186/s12906-023-04147-w

Hosmer, D.W., Lemeshow, S., Sturdivant, R.X. 2013. *Applied Logistic Regression*, vol. 398. Hoboken, NJ: Wiley.

Jităreanu, A., Trifan, A., Vieriu, M., Caba, I. C., Mârțu, I., Agoroaei, L. 2022. Current trends in toxicity assessment of herbal medicines: A narrative review. *Processes* 11(1): 83.

Jităreanu, A., Trifan, A., Vieriu, M., et al. 2023. Current trends in toxicity assessment of herbal medicines: A narrative review. *Processes* 11(1): 83. https://doi.org/10.3390/pr11010083

Li, A.P., He, Y.H., Zhang, S.Y., Shi, Y.P. 2022. Antibacterial activity and action mechanism of flavonoids against phytopathogenic bacteria. *Pesticide Biochemistry and Physiology* 188: 105221. https://doi.org/10.1016/j.pestbp.2022.105221

Li, P., Huang, C., Fu, Y., Wang, J., Wu, Z., Ru, J., ... Wang, Y. 2015. Large-scale exploration and analysis of drug combinations. *Bioinformatics* 31(12): 2007–2016.

Maroyi, A. 2012. Garden plants in Zimbabwe: Their ethnomedicinal uses and reported toxicity. *Ethnobotany Research and Applications* 10: 45–57.

Monfort-Lanzas, P., Gostner, J.M., Hackl, H. 2025. Modeling omics dose-response at the pathway level with DoseRider. *Computational and Structural Biotechnology Journal* 3(27): 1440–1448. https://doi.org/10.1016/j.csbj.2025.04.004. PMID: 40242291; PMCID: PMC12001094

Mongalo, N.I., McGaw, L.J., Finnie, J.F., et al. 2015. *Securidaca longipedunculata* Fresen (Polygalaceae): A review of its ethnomedicinal uses, phytochemistry, pharmacological properties and toxicology. *Journal of Ethnopharmacology* 165: 215–226. https://doi.org/10.1016/j.jep.2015.02.041

Mustapha, A.A. 2013. Ethno-medico-botanical uses of *Securidaca longepedunculata* Fresen (family-Polygalaceae) from Keffi Local Government, Nasarawa State, Nigeria. *Journal of Natural Remedies* 13(2): 133–137.

Ndamitso, M.M., Mohammed, A., Jimoh, T.O., et al. 2013. Phytochemical and antibacterial activity of *Securidaca longepedunculata* on selected pathogens. *African Journal of Microbiology Research* 7(50): 5652–5656. https://doi.org/10.5897/AJMR2013.5354

Ngonda, F., Magombo, Z., Mpeketula, P., et al. 2012. Evaluation of Malawian *Vernonia glabra* (Steetz) Vatke leaf and *Securidaca longepedunculata* (Fresen) root extracts for antimicrobial activities. *Journal of Applied Pharmaceutical Science* 2(11): 26–33. https://doi.org/10.7324/JAPS.2012.21106

Obi, C.L., Potgieter, N., Randima, L.P., et al. 2002. Antibacterial activities of five plants against some medically significant human bacteria. *South African Journal of Science* 98(1–2): 25–28.

Ochora, D.O., Kakudidi, E., Namukobe, J., et al. 2021. A new benzophenone, and the antiplasmodial activities of the constituents of *Securidaca longipedunculata* Fresen (Polygalaceae). *Natural Product Research* 0(0): 1–9. https://doi.org/10.1080/14786419.2021.1925272

Ochora, D.O., Kakudidi, E.K., Namukobe, J., et al. 2022. Synergism in antiplasmodial activities of artemether and lumefantrine in combination with *Securidaca longipedunculata*. *Plants* 11(47): 1–14.

Oyedeji-Amusa, M., Cuboia, N., Olofinsan, K. 2024. Medicinal plants used in the treatment of typhoid fever in Nigeria: A systematic review. *Journal of Herbal Medicine* 47(August): 100930. https://doi.org/10.1016/j.hermed.2024.100930

Plants of the World Online, POWO (2025). Distribution of *Securidaca longepedunculata*. https://powo.science.kew.org/taxon/urn:lsid:ipni.org:names:692714-1 (Accessed 5 January 2026).

Shai, K., Hassan, Z.B., Lebelo, S., et al. 2024. Chemical analysis and biological activities of various parts of *Securidaca longipedunculata* from South Africa. *Natural Product Communications* 19(9). https://doi.org/10.1177/1934578X241261018

Wang, J., Sheng, Z., Liu, Y., Chen, X., Wang, S., Yang, H. 2023. Combined proteomic and transcriptomic analysis of the antimicrobial mechanism of tannic acid against *Staphylococcus aureus*. *Front Pharmacol.* 14:1178177. https://doi.org/10.3389/fphar.2023.1178177

Wei, M., Chen, Q., Zhou, Y., Tie, H. 2025. The influence of synergistic antibacterial saponins, sapindoside A and B, on the fatty acid composition and membrane properties of *Micrococcus luteus*. *J Sci Food Agric.* 105(5):2777–2786. https://doi.org/10.1002/jsfa.14056

Chapter 9

Conservation and cultivation strategies of *Securidaca longepedunculata* Fresen

Dancan M Wakoli,[1] Duncan M Chalo,[2] Anywar Godwin U,[3] and Douglas Ongeri Ochora[4]

[1]Department of Biochemistry and Molecular Biology, Egerton University, Njoro, Kenya

[2]Department of Biological Sciences, Faculty of Science and Technology, University of Nairobi, Nairobi, Kenya

[3]Department of Plant Sciences, Microbiology and Biotechnology, College of Natural Sciences, Makerere University, Kampala, Uganda

Ethnopharmacology and Zoopharmacognosy, Bernhard Nocht Institute for Tropical Medicine, Hamburg, Germany

[4]Department of Biological Sciences, School of Pure and Applied Sciences, Kisii University, Kisii, Kenya

9.1 INTRODUCTION

Most people in Africa rely on medicinal plants as their first-line treatment for various diseases (Van Andel et al., 2015). Natural products are easily accessible and affordable, with fewer or no side effects. This realization has contributed to the increased demand for plant-based therapeutics in both developing and developed countries (Daboué et al., 2024). Trade in herbal products has increased to meet the demand. Consequently, there is a rising commercial exploitation of wild medicinal plant populations (Van Andel et al., 2015). Researchers have also focused on medicinal plants in search of novel chemical compounds with unique therapeutic activities (Madsen & Smith-Hall, 2023), therefore contributing to the overharvesting of medicinal plants.

Harvesting of wild products, especially trade in herbal medicine, acts as a source of income for more than 25% households in tropical and subtropical countries through the sale of plants, collection, transportation, and processing (Madsen & Smith-Hall, 2023; Van Andel et al., 2015). The depletion of wild environmental products is likely to shift harvesting from wild habitats to agricultural environments (Madsen & Smith-Hall, 2023). This calls for protection and increased cultivation of vulnerable plant species.

Correspondence: dancanwakoli@gmail.com

DOI: 10.1201/9781003544463-9

Among the most widely harvested plants is *Securidaca longepedunculata* Fresen, commonly called violet tree, belonging to the family Polygalaceae. *S. longepedunculata* is one of the two species of this genus occurring in Africa (Johnson, 1987). Despite it being endemic to the region, it has a wide distribution in tropical Africa and is restricted only to the North West and Limpopo provinces of South Africa. However, the populations of *S. longependuculata* face major conservation threats (Mukhithi, 2023).

The plant has a high cultural value based on its popularity as a medicinal plant. *S. longepedunculata* is considered a plant that can treat every conceivable ailment (Ouédraogo et al., 2017; Tikisa et al., 2019). This has made the plant vulnerable to exploitation, leading to its overharvesting. Roots and stems are the most widely used parts of *S. longepedunculata*, which contribute to its vulnerability to extinction (Kloos, 2023; Mulumbati et al., 2023).

A high demand for *S. longepedunculata* has contributed to its exploitation in various African countries (Mulumbati et al., 2023). Other factors, such as climate change, have also contributed to the decreased wild populations of *S. longepedunculata* in the Sahel region (Nadembega et al., 2011). Kaniaru et al. (2021) also report threats in firewood extraction. Therefore, there is an urgent need for the conservation of *S. longepedunculata* for its sustainable utilization. This will contribute towards a large-scale and year-round supply of plant materials to meet the high demand for herbal medicine (Kloos, 2023).

Reforestation is recommended as a means of overcoming these challenges. However, natural cultivation of *S. longepedunculata* is challenged by seed dormancy, low seed germination, high seedling mortality, low seedling growth, and easily broken taproots (Daboué et al., 2024). Manual seed collection is also limited by the decreased number of trees, decreased fruiting, and can result in ecosystem disruption (Mulumbati et al., 2023). Another challenge has been reported through phenological studies of *S. longepedunculata*, where hailstorms destroy flowers, leading to decreased fruiting (Kaniaru et al., 2021). To better understand ways of overcoming these challenges, this chapter discusses the conservation and cultivation of *S. longepedunculata*.

9.2 METHODS

A scientific literature search was conducted on Scopus, PubMed, and Google Scholar databases. The search strategy employed the use of specific subject headings and keywords, including "cultivation of *Securidaca*," "conservation of *Securidaca*," and "conservation of herbal medicine." Only publications and abstracts written in English were included in the review. No restrictions were placed on the year of publication or author affiliation during the selection process.

9.3 DISCUSSION

9.3.1 Taxonomy of *Securidaca longepedunculata*

9.3.1.1 *Is it* Securidaca longepedunculata *Fresen or* Securidaca longepedunculata *Fresen?*

You need to know what you are protecting, and you cannot conserve what you don't know. People love and protect what they know. Indeed, accurate plant taxonomy is a "sin que non" to meaningful conservation (Johnson, 1987). The scenario of discrepancy in the names—*Securidaca longepedunculata* Fresen and *Securidaca longepedunculata* Fresen—is of interest to note regarding the great socio-economic value of this plant. From Global Biodiversity Information Facility (GBIF, 2025), occurrences of collection records of both names are at 2,092 for *S. longepedunculata* and at 2,899 for *S. longepedunculata*. According to Plant of the World Online (POWO, 2025) and International Plant Names Index (IPNI, 2025), the holotype specimen (the original/reference used by the author in the description of a new specimen) was collected from Ethiopia and deposited at Museum Senckenbergianum in 1837. The species name was first published in Mus. Senckenberg. 2: 275 (1837) as *Securidaca longepedunculata* Fresen (GBIF, 2025). Over the years, eight heterotypic synonyms of the species have been adopted, namely, *Lophostylis angustifolia* Hochst in Flora 25: 231 (1842), *Lophostylis oblongifolia* Hochst in Flora 25: 231 (1842), *Lophostylis pallida* Klotzsch in W.C.H.Peters, Naturw. Reise Mozambique 6 (Bot., 1): 115 (1861), *Securidaca angustifolia* Miq. in Tijdschr. Wis- Natuurk. Wetensch. Eerste Kl. Kon. Ned. Inst. Wetensch. 1: 152 (1848), *Securidaca longepedunculata* var. angustifolia Robyns in Bull. Soc. Roy. Bot. Belgique 60: 89 (1927), *Securidaca longepedunculata* var. parvifolia Oliv. in Fl. Trop. Afr. 1: 134 (1868), *Securidaca oblongifolia* Benth. & Hook.f. ex Sond. in W.H.Harvey & auct. suc. (eds.), Fl. Cap. 2: 586 (1862), *Securidaca spinosa* Sim in Forest Fl. Port. E. Afr.: 14 (1909), and also *Elsota longepedunculata* (Fresen) Kuntze in Revis. Gen. Pl. 1: 46 (1891) as a homotypic synonym (Flowering Plants of Africa, 2025).

Currently, *S. longepedunculata* Fresen is the accepted published taxon based on POWO (2025) and World Flora Online (WFO, 2025), and this is traceable to the description and in the naming of the holotype (GBIF, 2025).

9.3.2 Increased demand for *Securidaca longepedunculata*

The successful discovery of modern antimalarial drugs, quinine from the *Cinchona* tree and artemisinin from *Artemisia annua*, has reignited interest in medicinal plants as potential sources for new pharmaceuticals (Igoli et al., 2005). Across Africa, approximately 80% of the population depends on traditional medicine as their main source of primary healthcare (*African Traditional Medicine Day*, 2022).

S. longepedunculata, a native African medicinal plant, is the most popular plant used for traditional medicine in most African countries (POWO, 2025). Hence, its demand remains high. *S. longepedunculata* is used for both conventional and contemporary medicine in most African countries (Mongalo et al., 2015; Ojewole, 2008). It is thought to have more than one hundred medicinal applications (Dapar et al., 2007). It has multiple bioactive compounds in the bark, roots, and leaves that are used for treating multiple ailments, namely, malaria, stomach ache, toothache, headache, sleeping sickness, cough, epilepsy, chest complaints, asthma, cancer, snakebites, wound dressing, diarrhea, gout, boils, gonorrhea, fever, abortion, ritual suicide, tuberculosis, infertility, constipation, prevention of breastfeeding infant illnesses, diabetes, managing opportunistic diseases associated with HIV/AIDS, and for placental removal (Ajali & Chukwurah, 2004; Anywar et al., 2021; Mudau et al., 2022; Ochora, Kakudidi, Namukobe, Heydenreich, et al., 2022; Ochora, Kakudidi, Namukobe, Ipulet, et al., 2022; Suzana Augustino, 2011). Additionally, the active ingredients from its bark can be used for treating convulsions in children, rheumatic and inflammatory conditions, microbial infections, managing hypertension and paralysis (Ajali & Chukwurah, 2004).

Despite its prominent role in traditional African medicine, the properties of *S. longepedunculata* extend beyond therapeutic use. A combination of its methanolic root extracts and methyl salicylate components can be formulated into a potent poison, traditionally utilized in various West African countries. This natural toxin has been used for hunting, defense against wild animal attacks, and as a weapon in tribal warfare (Neuwinger, 1998). Moreover, it serves as an effective, low-cost pesticide for protecting stored grains such as maize from weevils and other insect pests, offering an accessible alternative to synthetic chemical treatments (Stevenson et al., 2014).

Beyond its medicinal and pesticidal applications, *S. longepedunculata* roots and ground seeds are used in the production of soap and bleaching agents, while the leaves are consumed as vegetables in some communities. In countries like Eritrea, the plant is also valued as a vital source of nectar for honey production (Booth & Wickens, 1988). It can also be harvested for timber, firewood, burning charcoal, and for high-quality fiber that can be used for making textiles, string, and rope (Booth & Wickens, 1988). Due to its resistance to rot and termite damage, the wood is exclusively used for hut construction, arrow making, and broom crafting across Western, Central, and Southern Africa (Van Wyk & Gericke, 2000).

These diverse applications contribute to the continued and growing demand for *S. longepedunculata*, particularly its roots (Tiawoun et al., 2017). However, this high demand poses a significant threat to the species' survival. Natural regeneration is limited by several factors, including high seed dormancy, low germination rates, slow growth, and the plant's brittle taproot, which complicates transplantation. Furthermore, intensive harvesting of the roots has led to overexploitation, placing additional pressure on wild populations (Lijalem & Feyissa, 2020; Mbuya et al., 1994).

9.3.3 Overexploitation and conservation of *Securidaca longepedunculata*

Quantitative market surveys in West Africa (Benin and Ghana) depict *S. longepedunculata* as a vulnerable plant due to its overharvesting. Market vendors consider *S. longepedunculata* as a scarcely available plant (Van Andel et al., 2015). In Burkina Faso, *S. longepedunculata* has become rare or disappeared locally among the ligneous plants of the country (Ouédraogo et al., 2017). Also, the growing urbanization in the country has inevitably led to the excessive harvesting of medicinal plants in the forests surrounding large cities, resulting in their overexploitation (Ouedraogo et al., 2017). Similarly, in Kano State, Nigeria, *S. longepedunculata* is among the most frequently mentioned threatened medicinal plants by traditional practitioners (Abubakar et al., 2018). Ugandans consider it a rare medicinal plant whose availability is decreasing. The plant was not found in any of the 320 plots surveyed in Nawaikoke sub-county in Uganda (Tabuti et al., 2011). In Limpopo Province, South Africa, *S. longepedunculata* is among the plants of conservation concern accorded with status such as critically endangered, declining, vulnerable, threatened, and data deficient (Semenya et al., 2013). *S. longepedunculata* has been overexploited in Lubumbashi, DR Congo, since the roots are highly valued in herbal medicine (Mulumbati et al., 2023). Furthermore, harvesters and vendors in Ghana consider the shredded roots of *S. longepedunculata* scarce and expensive, sold at a higher price ($47.8/kg) since it is considered rare and transported from a long distance (Van Andel et al., 2012). The plant is therefore sold in bulk due to the increased demand in the country (Van Andel et al., 2012). Therefore, conservation and cultivation of *S. longepedunculata* is warranted in order to meet the growing demand and to prevent overexploitation of this endangered species (Joshi & Joshi, 2014).

The conservation status of this plant is also of great concern since the roots are the most widely used part, thereby exposing the plant to the risk of eradication (Nguta, 2019). Reports in Ethiopia indicate that the species is prone to extinction due to overexploitation for its medicinal properties (Kloos, 2023; Mesfin et al., 2009; Tewelde & Mesfin, 2020). In South Africa, the species is red-listed for conservation (Zulu et al., 2011. The species is also among the most highly exploited in the Guinean Savannah Highlands in Northern Cameroon (Mapongmetsem et al., 2012). In Ghana and Benin, *S. longepedunculata* is reported to be in high commercial demand in markets and becoming increasingly scarce in the wild (Abubakar et al., 2018; Van Andel et al., 2015). Ouedraogo et al. (2020) report that the markets in Burkina Faso are selling *S. longepedunculata* at high prices mainly due to its rarity and recommend conservation measures to preserve the natural populations. Efforts to conserve the tree have been initiated by global entities such as the Kew Botanical Garden (Liu et al., 2018). In Kenya, conservation of the threatened population of *S. longependuculata* at Makueni county using the local community has been successful (Kaniaru et al., 2021). Furthermore, sustainable root harvesting techniques by the local communities are encouraged (Nguta, 2019). Studies on sexual and asexual reproductions using seeds, branches, and root suckers under various conditions have shown to be a solution toward the conservation of *S. longependuculata* (Tiawoun et al., 2015). Recent studies on vegetation propagation on root cuttings have been reported (Lijalem & Feyissa, 2020).

9.3.4 Cultivation of *Securidaca longepedunculata*

S. longepedunculata, commonly known as the violet tree, exhibits specific requirements for seed germination, soil conditions, and environmental factors that influence its growth. Understanding these aspects is crucial for effective propagation and conservation efforts. Although not cultivated, different authors such as Lijalem and Feyissa (2020) and Mulumbati et al. (2023) have established different micropropagation techniques that, if adopted and used on a large scale, can support the cultivation of *S. longepedunculata*.

S. longepedunculata is a fairly fast-growing tree and can also grow from root suckers (Katende et al., 1995). Propagation of *S. longepedunculata* is primarily achieved through seeds and vegetative materials, using either *ex vitro* (natural) or *in vitro* (artificial) techniques. Among these, *in vitro* propagation methods have demonstrated greater effectiveness, offering enhanced seed germination, shoot multiplication, and root development.

9.3.4.1 Seed germination

S. longepedunculata seeds have a very low and erratic germination under natural conditions (Tiawoun et al., 2015). Seeds germinate with difficulty if not pretreated. Germination of treated seeds is good and fairly fast after soaking in cold water for 24 hours (Katende et al., 1995). The seeds can retain viability for a long period at room temperature if kept dry (Katende et al., 1995). Optimal germination occurs under controlled conditions, with *in vitro* methods achieving rates of 67–90%, while *ex vitro* methods yield only 43% (Zulu et al., 2011). These findings highlight the potential of *in vitro* propagation as a more reliable and efficient approach for the conservation and mass production of *S. longepedunculata*.

9.3.4.2 Ex vitro *propagation*

The seeds of *S. longepedunculata* exhibit prolonged dormancy, often lasting up to one year, necessitating a delayed planting period. However, despite this extended dormancy, the seeds tend to lose viability quickly, and their germination remains highly unpredictable (Lijalem & Feyissa, 2020). Traditional propagation practices involve soaking the seeds in cold water for 24 hours, followed by planting them in soil and covering them with water and grass until the onset of rainfall. However, it is not suitable for large-scale propagation due to its low germination success. Even when germination occurs, seedling development is slow, and the fragile taproot system poses significant challenges for transplantation (Mbuya et al., 1994). Given these limitations, *in vitro* propagation techniques are strongly recommended for the large-scale cultivation of *S. longepedunculata*, whether in agroforestry systems or for reintroduction into natural habitats. These methods offer a more controlled and efficient approach to overcoming dormancy issues, enhancing germination rates, and supporting healthier root and shoot development (Lijalem & Feyissa, 2020).

9.3.4.3 **In vitro** *propagation*

In vitro propagation, also known as micropropagation, is a plant cultivation technique that involves the use of plant parts or tissues (explants) grown in a controlled, sterile environment on an artificial nutrient medium. This method allows the regeneration of whole plants from explants such as shoots, root tips, callus tissue, seeds, embryos, pollen grains, ovules, or even single cells (Kaya & Huyop, 2020; Vidyagina et al., 2021). It is particularly advantageous for the rapid multiplication of plant species that are difficult to propagate through conventional methods, or those facing threats of extinction, as it effectively bypasses seed dormancy and viability issues.

In the case of *S. longepedunculata*, *in vitro* propagation presents a critical solution for both conservation and cultivation. The technique facilitates mass production of this high-demand species, ensuring a sustainable supply while reducing pressure on wild populations. Several African countries where *S. longepedunculata* is native have developed *in vitro* propagation protocols to support its preservation and wider use in traditional and modern medicine (Lijalem & Feyissa, 2020; Mulumbati et al., 2023; Zulu et al., 2011). *S. longepedunculata* can be cultivated using the seeds or shoot cuttings or tissue cultures.

9.3.4.3.1 In vitro *propagation using seeds*

The seeds of *S. longepedunculata* are enclosed in a hard seed coat, which contributes to their prolonged dormancy, often lasting up to one year, and low germination rates under natural conditions (Lijalem & Feyissa, 2020). In the context of *in vitro* propagation, several seed pretreatment techniques have been developed to overcome the dormancy barriers and improve germination and seedling establishment. These methods are designed to weaken or break the hard seed coat, thereby enhancing water and gas permeability and triggering germination (Daboué et al., 2024).

Scarification is one such technique used to physically or chemically disrupt the seed coat. Physical scarification involves mechanically nicking, cutting, or rubbing the seed surface using rough objects such as sandpaper or files, or by vigorously shaking seeds with abrasive materials. Thermal methods, such as exposing seeds to hot water, may also be employed. Chemical scarification typically involves soaking seeds in concentrated sulfuric acid for a specified duration to erode the seed coat. Stratification, another common method, mimics natural environmental cues by subjecting seeds to moist, cold conditions. This often includes soaking the seeds in water for 1–2 days, followed by storage in a humid medium at low temperatures (3–5 °C) for a predetermined period (Nelson, 2015). This process is especially effective for species adapted to temperate or seasonal climates, where cold periods naturally break dormancy. These pretreatment techniques, when integrated into *in vitro* propagation protocols, significantly improve the germination success of *S. longepedunculata*, offering a viable strategy for its mass propagation.

Several pretreatment methods have been employed to overcome the prolonged seed dormancy of *S. longepedunculata*, thereby enhancing germination under laboratory, greenhouse, and field conditions (Lijalem & Feyissa, 2020; Mulumbati et al., 2023; Tiawoun et al., 2017; Zulu et al., 2011). These methods include mechanical and chemical approaches, such as seed coat removal, treatment with concentrated sulfuric acid, gibberellic acid, hot and cold water, Clorox, 10% sodium hypochlorite, and 70% ethanol. Other techniques involve the application of cytokinins, sowing seeds on moistened filter paper, and cutting the seeds transversely to expose the embryo and facilitate germination (Lijalem & Feyissa, 2020; Mulumbati et al., 2023; Tiawoun et al., 2017; Zulu et al., 2011).

Investigations of *S. longepedunculata* seed germination under laboratory, greenhouse, and field conditions have demonstrated that various pretreatment methods can significantly influence germination success. Among the methods tested—seed coat removal and treatments with sulfuric acid, gibberellic acid, hot water, and cold water—seed coat removal emerged as the most effective technique. It resulted in germination rates of 90% in the greenhouse and 63% in the laboratory (50). Soaking seeds in 95–97% sulfuric acid for 10 minutes achieved a high germination rate of 82.3%. However, longer treatment duration beyond 10 minutes led to a decline in germination, suggesting that prolonged acid exposure may damage the seed embryo, thereby reducing viability (Wang et al., 2007). Similarly, treatment with gibberellic acid was shown to enhance both germination and seedling growth. The highest germination rate of 70% was recorded at a concentration of 400 mg/L (Tiawoun et al., 2017). Cold water soaking for 48 hours produced a germination rate of 46.7% and was also associated with a higher seedling survival rate. In contrast, soaking seeds in boiled water failed to improve germination, likely due to embryo damage caused by high temperatures (El Nour et al., 1993; Tiawoun et al., 2017). Despite these findings, there are conflicting reports regarding the effect of hot water treatment on *S. longepedunculata* seed germination, indicating the need for further investigation into optimal treatment conditions (El Nour et al., 1993; Tiawoun et al., 2017).

A study conducted in Ethiopia investigated the effects of Clorox treatment, seed de-coating, and transverse cutting of the seed tip on the germination of *S. longepedunculata*. The results demonstrated that treating seeds with Clorox for 10 minutes yielded a high decontamination rate (85%) and a germination rate of 80%. However, extending the treatment duration to 15 minutes led to a decline in germination, likely due to reduced seed viability caused by overexposure to the chemical (Lijalem & Feyissa, 2020). Decoating the seeds significantly improved germination compared to seeds with intact seed coats, which showed no germination. This enhancement is attributed to improved permeability, allowing better penetration of water and gases necessary for initiating the germination process (Lijalem & Feyissa, 2020; Zulu et al., 2011). Interestingly, other studies have reported that coated seeds can germinate when treated with gibberellins, suggesting that chemical stimulation may overcome the physical dormancy (Zulu et al., 2011). Additionally, transverse cutting at the tip of de-coated seeds resulted in 100% germination on Murashige and Skoog (MS) medium, within 7 to 10 days. This technique appears to enhance germination by reducing seed size, thereby enabling the embryo to more efficiently utilize the limited moisture and nutrients available in the medium (Lijalem & Feyissa, 2020).

Successful *in vitro* propagation of *S. longepedunculata* depends largely on maintaining sterile conditions throughout the process. As such, pretreatment of seeds with disinfectants is a critical step to eliminate microbial contaminants that could compromise germination and seedling development (Mastuti et al., 2017; Silva et al., 2015). However, while disinfection is necessary, excessive exposure to sterilizing agents can be toxic, leading to reduced seed viability and impaired germination (Lijalem & Feyissa, 2020; Mulumbati et al., 2023). A protocol developed in the DR Congo demonstrated that treating *S. longepedunculata* seeds with 10% sodium hypochlorite for 10 minutes, followed by 70% ethanol for 5 minutes, significantly reduced microbial contamination and enhanced germination rates, reaching up to 77% (Mulumbati et al., 2023). Nevertheless, an extended soaking period, although effective in further minimizing contamination, was found to negatively impact germination, likely due to chemical damage to the seed tissues. This highlights the importance of optimizing sterilization protocols to balance decontamination efficiency with seed viability.

The growth and development of plantlets from germinated *S. longepedunculata* seeds are significantly influenced by plant hormones, particularly cytokinins and auxins. The incorporation of these growth regulators into culture media has been shown to enhance propagation outcomes. Specifically, the addition of cytokinins promotes the development of axillary shoots, thereby increasing the number of plantlets produced from a single germinated seed (Mulumbati et al., 2023; Tevfik & Yegorova, 2020). *In vitro* cultivation of *S. longepedunculata* on media supplemented with 1.5 mg/L of cytokinin compounds, 6-benzylaminopurine (BAP), kinetin, and meta-topolin has demonstrated a significant increase in plantlet production. Among these, meta-topolin yielded the highest number of plantlets, outperforming both BAP and kinetin. In contrast, seeds germinated on hormone-free media produced only a single plantlet. These findings underscore the importance of hormonal supplementation in optimizing micropropagation efficiency.

The effectiveness of plant growth regulators is highly dependent on both the type of hormone and its concentration. Research indicates that lower concentrations of cytokinins stimulate the activation of mitogen-activated protein kinases (MAPKs), which are critical regulators of cellular growth and differentiation. However, higher concentrations of cytokinins can inhibit MAPK activity, thereby limiting growth (Carra et al., 2012). This suggests that achieving an optimal hormonal balance is essential for maximizing plantlet development and propagation success.

9.3.4.3.2 In vitro *propagation of* S. longepedunculata *using vegetative parts*

Vegetative propagation is a method of plant cultivation that involves using the vegetative parts of the parent plant, such as leaves, stems, or roots, to develop new, genetically identical plants. This technique is particularly valuable for the propagation of medicinal plants, as it ensures the preservation of desirable genetic traits and medicinal properties (Zéphirin et al., 2019). *S. longepedunculata* can be effectively propagated on a large scale using this method. In both Ethiopia and the DR Congo, this approach has primarily utilized shoot segments for plant regeneration (Lijalem & Feyissa, 2020; Mulumbati et al., 2023). The cultivation process typically involves three key stages: shoot initiation, shoot multiplication, and rooting. These stages are carefully managed to ensure optimal growth and establishment of healthy plantlets capable of being transplanted into natural or cultivated environments (Mulumbati et al., 2023). Vegetative propagation, therefore, offers a reliable and efficient method for the large-scale cultivation and conservation of *S. longepedunculata*, particularly in response to increasing demand and the need to protect wild populations.

Shoot initiation is the first critical stage in the vegetative propagation of *S. longepedunculata*, involving the preparation of shoot segments from the parent plant. In this process, shoots are cut into approximately 1 cm segments and cultured on a shoot initiation medium, typically MS medium supplemented with plant growth regulators such as BAP, thidiazuron (TDZ), or a combination of both. The addition of these cytokinins to the culture medium stimulates shoot induction, with visible initiation usually occurring within one week.

The response of explants to hormonal treatments varies depending on their endogenous levels of growth regulators. In a protocol developed for the *in vitro* propagation of *S. longepedunculata*, supplementation with 1.0 mg/L BAP alone yielded the highest shoot initiation rate of 80%. A combination of 1.0 mg/L BAP and 0.5 mg/L TDZ resulted in a slightly lower initiation rate of 67%. However, the use of TDZ alone, or in combination with BAP at other concentrations, did not achieve comparable initiation efficiency. These findings highlight BAP at 1.0 mg/L as the most effective hormone treatment for initiating shoot development in *S. longepedunculata* (Mulumbati et al., 2023).

The shoot multiplication stage follows successful shoot initiation and involves culturing the initiated shoots in a multiplication medium to induce the formation of multiple shoots. This stage typically employs MS medium supplemented with a combination of cytokinins and low concentrations of auxins. Specifically, BAP combined with indole-3-butyric acid (IBA), or TDZ combined with α-naphthalene acetic acid (NAA), is commonly used. Studies have shown that the synergistic effect of cytokinins and low levels of auxins significantly enhances shoot proliferation, producing a higher number of shoots per explant (Tsay et al., 1989). In a comparative study, the use of 1.5 mg/L BAP in combination with 0.1 mg/L IBA produced a greater number of shoots with optimal length, highlighting the effectiveness of BAP in promoting shoot multiplication (Bramhanapalli et al., 2017). In contrast, shoot cultures grown on media containing TDZ in combination with NAA resulted in shorter and fewer shoots, many of which were dwarfed. The least favorable outcome was observed in control media lacking any growth regulators, which produced the fewest and shortest shoots (Mulumbati et al., 2023).

The rooting stage is a critical phase in the micropropagation of *S. longepedunculata*, during which shoots are transferred to rooting media supplemented with auxins to induce root formation. Typically, shoots are cultured in the dark for 4–5 weeks on MS medium enriched with auxins such as indole-3-acetic acid (IAA), IBA, or NAA. These plant growth regulators play a pivotal role in promoting both root induction and elongation. Among the concentrations tested, 2.0 mg/L IAA produced the highest number of roots, while 3.0 mg/L IAA yielded the longest roots. In contrast, no rooting occurred in media lacking auxins, underscoring their essential role in the process (Mulumbati et al., 2023).

Root development from microcuttings remains a complex and often limiting step in *in vitro* propagation, requiring precise optimization of conditions to ensure success (Erst et al., 2018; Fogaça & Fett-Neto, 2005). In *S. longepedunculata*, the rooting and recovery efficiency of apical, median, and basal microcuttings varies according to the type and concentration of auxins used. Basal microcuttings tend to show the highest recovery and rooting success, likely due to their greater physiological maturity and enhanced tolerance to abiotic stressors such as temperature fluctuations (De Klerk, 2002; Henselová et al., 2002). This physiological robustness contributes to faster and more consistent root formation (De Klerk, 2002). Interestingly, unlike many other plant species that show a distinct preference for specific auxins, *S. longepedunculata* does not exhibit a strong bias toward any particular auxin type, allowing for greater flexibility in choosing effective rooting treatments (Lijalem & Feyissa, 2020).

9.3.5 Soil and environmental conditions

S. longepedunculata thrives in well-drained soils typical of savanna ecosystems, where it co-dominates. Soil composition and moisture levels significantly affect seedling establishment, as the species is sensitive to high-intensity fires that can damage root systems (Tshisikhawe et al., 2012).

9.4 CONCLUSION

The wide use of *S. longepedunculata* as a medicinal plant has led to its excessive harvesting to meet the increased demand, especially for the roots. This overexploitation requires urgent conservation of the plant species. However, natural regeneration of *S. longepedunculata* is challenged by various factors, such as seed dormancy, low seed germination, and low growth rate.

Approaches such as *in vitro* and *ex vitro* propagation of different parts of *S. longepedunculata* have shown potential for rapid multiplication of the plant in an artificial medium under sterile conditions. This will promote the conservation of *S. longepedunculata,* ensuring a sustainable, year-round supply of plant materials to meet the high demand for herbal medicine.

REFERENCES

About the World Flora Online Project WFO's June 2025 release | World Flora Online. Available at: https://about.worldfloraonline.org/june-2025-release (accessed 17 September 2025).

Abubakar, U. S., Khalifa, B. I., Abdu, F., Gawuna, T. A., Adamu, J. G., & Rogo, S. S. (2018). Threatened medicinal plants of kano flora and the need for urgent conservation. *International Journal of Conservation Science*, *9*(1), 173–178.

Ajali, U., & Chukwurah, B. K. C. (2004). Antimicrobial activity of *Securidaca longipedunculata. Phytomedicine*, *11*(7), 701–703. https://doi.org/10.1016/j.phymed.2003.07.002

Anywar, G., Kakudidi, E., Byamukama, R., Mukonzo, J., Schubert, A., Oryem-Origa, H., & Jassoy, C. (2021). A review of the toxicity and phytochemistry of medicinal plant species used by herbalists in treating people living with HIV/AIDS in Uganda. *Frontiers in Pharmacology*, *12*. https://doi.org/10.3389/fphar.2021.615147

Booth, F. E. M., & Wickens, G. E. (1988). *Non-timber Uses of Selected Arid Zone Trees and Shrubs in Africa*. Food & Agriculture Org.

Bramhanapalli, M., Thogatabalija, L., & Gudipalli, P. (2017). Efficient in vitro plant regeneration from seedling-derived explants and genetic stability analysis of regenerated plants of Simarouba glauca DC. by RAPD and ISSR markers. *In Vitro Cellular & Developmental Biology – Plant*, *53*(1), 50–63. https://doi.org/10.1007/s11627-016-9795-0

Carra, A., Sajeva, M., Abbate, L., Siragusa, M., Sottile, F., & Carimi, F. (2012). In vitro plant regeneration of caper (Capparis spinosa L.) from floral explants and genetic stability of regenerants. *Plant Cell, Tissue and Organ Culture (PCTOC)*, *109*(2), 373–381. https://doi.org/10.1007/s11240-011-0102-9

Daboué, E. M. S., Béné, A., Dimobe, K., Dayamba, D. S., Zouré, A. B., Ouattara, B., Sabo, P., Tuina, S., Neya, O., Vinceti, B., & Ouédraogo, A. (2024). Variability in fruit morphology and germination capacity of the tropical medicinal species Securidaca longipedunculata Fres. *Seeds*, *3*(4), Article 4. https://doi.org/10.3390/seeds3040042

Dapar, M., Aguiyi, J., Wannang, N., Gyang, S., & Tanko, N. M. (2007). The histopathologic effects of Securidaca longepedunculata on heart, liver, kidney and lungs of rats. *African Journal of Biotechnology*, (ISSN: 1684-5315) 6(5), 6.

da Silva, J. A. T., Winarto, B., Dobránszki, J., & Zeng, S. (2015). Disinfection procedures for *in vitro* propagation of *Anthurium*. *Folia Horticulturae*, *27*(1), 3–14. https://doi.org/10.1515/fhort-2015-0009

De Klerk, G.-J. (2002). Rooting of microcuttings: Theory and practice. *In Vitro Cellular & Developmental Biology – Plant*, *38*(5), 415–422. https://doi.org/10.1079/IVP2002335

El Nour, M. F., El Khalifa, K., Massimo, K., & El Hassen, B. (1993). Preliminary study on seed pregermination treatment and vegetative propagation of Balanites aegyptiaca (L.) Del. https://agris.fao.org/search/en/providers/122439/records/6471e9c677fd37171a70e0fI

Erst, A. A., Gorbunov, A. B., & Erst, A. S. (2018). Effect of concentration, method of auxin application and cultivation conditions on in vitro rooting of bog blueberry (Vaccinium uliginosum L.). *Journal of Berry Research*, *8*(1), 41–53. https://doi.org/10.3233/JBR-170253

Fogaça, C. M., & Fett-Neto, A. G. (2005). Role of auxin and its modulators in the adventitious rooting of Eucalyptus species differing in recalcitrance. *Plant Growth Regulation*, *45*(1), 1–10. https://doi.org/10.1007/s10725-004-6547-7

Global Biodiversity Information Facility, GBIF. (2025). *The holotype specimen of Securidaca longepedunculata Senckenberg*. Herbarium Senckenbergianum (FR). Occurrence dataset. https://doi.org/10.15468/ucmdjy and www.gbif.org/occurrence/3046121603 (accessed 21 July 2025).

Henselová, M., Lux, A., & Masarovičová, E. (2002). Effect of growth regulators on rooting cuttings of Karwinskia species under in vivo conditions. *Plant, Soil and Environment*, *48*(10), 471–476. https://doi.org/10.17221/4397-pse

Igoli, J. O., Ogaji, O. G., Tor-Ayiin, T. A., & Igoli, N. P. (2005). Traditional medicine practice amongst the IGEDE people of Nigeria. Part II. *African Journal of Traditional, Complementary and Alternative Medicines*, 2(2), Article 2.

IPNI. (2025). *International Plant Names Index*. Royal Botanic Gardens, Kew; Harvard University Herbaria & Libraries; Australian National Herbarium. https://www.ipni.org [Retrieved 17 October 2025].

Johnson, C. T. (1987). Taxonomy of the African species of Securidaca (Polygalaceae). *South African Journal of Botany*, *53*(1), 5–11.

Joshi, B. T., & Joshi, R. K. (2014). The role of medicinal plants in livelihood improvement in Uttarakhand. *International Journal of Herbal Medicine*, *1*(6), 55–58.

Kaniaru, K., Wafula, C. M., Sila, J., & Alindalwanyi, C. (2021). *Community-Based Conservation of Locally Threatened, Medicinal Securidaca longepedunculata Fresen (Violet Plant); For the People, by the People in the Drylands of Kenya* (pp. 1–93). Nairobi National Museum.

Katende, A. B., Birnie, A., & Tengnas, B. (1995). Useful trees and shrubs for Uganda. *Identification, propagation and management for agricultural and pastoral communities. Regional soil conservation unit (RSCU), Swedish International Development Authority (SIDA)*, 1–710.

Kaya, Y., & Huyop, F. Z. (2020). An easy and reliable method for establishment and maintenance of tissue cultures of Nicotiana tabacum cv TAPM 26. *International Journal of Science Letters*, *2*(2), Article 2. https://doi.org/10.38058/ijsl.764947

Kloos, H. (2023). Challenges and prospects of medicinal plant sustainability in Ethiopia. *Journal of Pharmacy and Pharmacology Research*, *7*(4), 233–242.

Lijalem, T., & Feyissa, T. (2020). In vitro propagation of Securidaca longipedunculata (Fresen) from shoot tip: An endangered medicinal plant. *Journal of Genetic Engineering and Biotechnology*, *18*(1), 3. https://doi.org/10.1186/s43141-019-0017-0

Liu, U., Breman, E., Cossu, T. A., & Kenney, S. (2018). The conservation value of germplasm stored at the millennium seed bank, royal botanic gardens, Kew, UK. *Biodiversity and Conservation*, *27*(6), 1347–1386.

Madsen, S. T., & Smith-Hall, C. (2023). Wild harvesting or cultivation of commercial environmental products: A theoretical model and its application to medicinal plants. *Ecological Economics*, *205*, 107701. https://doi.org/10.1016/j.ecolecon.2022.107701

Mapongmetsem, P. M., Kapchie, V. N., & Tefempa, B. H. (2012). Diversity of local fruit trees and their contribution in sustaining the rural livelihood in the Northern Cameroon. *Ethiopian Journal of Environmental Studies and Management*, *5*(1). https://doi.org/10.4314/ejesm.v5i1.5

Mastuti, R., Munawarti, A., & Firdiana, E. R. (2017). The combination effect of auxin and cytokinin on in vitro callus formation of Physalis angulata L. – A medicinal plant. *AIP Conference Proceedings*, *1908*(1), 040007. https://doi.org/10.1063/1.5012721

Mbuya, L. P., Msanga, H. P., Ruffo, C. K., Birnie, A., & Tengnas, B.,. (1994). *Useful trees and shrubs for Tanzania: Identification, propagation and management for agricultural and pastoral communities (No. 6.* xx+-542*)*.https://agris.fao.org/search/en/providers/122621/records/647396a03ed73003714cd8a8

Mesfin, F., Demissew, S., & Teklehaymanot, T. (2009). An ethnobotanical study of medicinal plants in Wonago Woreda, SNNPR, Ethiopia. *Journal of Ethnobiology and Ethnomedicine*, *5*(1), 28. https://doi.org/10.1186/1746-4269-5-28

Mongalo, N. I., McGaw, L. J., Finnie, J. F., & Staden, J. V. (2015). *Securidaca longipedunculata* Fresen (Polygalaceae): A review of its ethnomedicinal uses, phytochemistry, pharmacological properties and toxicology. *Journal of Ethnopharmacology*, *165*, 215–226. https://doi.org/10.1016/j.jep.2015.02.041

Mudau, T. E., Olowoyo, J. O., & Amoo, S. O. (2022). Ethnobotanical assessment of medicinal plants used traditionally for treating diabetes in Vhembe district, Limpopo Province, South Africa. *South African Journal of Botany*, *146*, 304–324. https://doi.org/10.1016/j.sajb.2021.10.016

Mukhithi, A. G. (2023). *The phenological study of Securidaca longepedunculata, in the Nylsvley Nature Reserve, Limpopo Province, South Africa* [PhD Thesis]. https://univendspace.univen.ac.za/bitstream/handle/11602/2636/Dissertation%20-%20Mukhithi%2C%20a.%20g.-.pdf?sequence=1&isAllowed=n

Mulumbati, M. C., Godoy Jara, M., Baboy Longanza, L., Bogaert, J., Werbrouck, S., Sikuzani, Y. U., & Mazinga Kwey, M. (2023). In vitro regeneration protocol for Securidaca longepedunculata Fresen, a threatened medicinal plant within the region of Lubumbashi (Democratic Republic of the Congo). *Conservation*, *3*(3), Article 3. https://doi.org/10.3390/conservation3030028

Nadembega, P., Boussim, J. I., Nikiema, J. B., Poli, F., & Antognoni, F. (2011). Medicinal plants in Baskoure, Kourittenga Province, Burkina Faso: An ethnobotanical study. *Journal of Ethnopharmacology*, *133*(2), 378–395. https://doi.org/10.1016/j.jep.2010.10.010

Nelson, S. O. (2015). Chapter 5—Seed treatment applications. In S. O. Nelson (Ed.), *Dielectric Properties of Agricultural Materials and their Applications* (pp. 57–71). Academic Press. https://doi.org/10.1016/B978-0-12-802305-1.00005-1

Neuwinger, H. D. (1998). Alkaloids in arrow poisons. In M. F. Roberts & M. Wink (Eds.), *Alkaloids: Biochemistry, Ecology, and Medicinal Applications* (pp. 45–84). Springer US. https://doi.org/10.1007/978-1-4757-2905-4_3

Nguta, J. M. (2019). In vivo antimalarial activity, toxicity, and phytochemical composition of total extracts from Securidaca longepedunculata Fresen (Polygalaceae). *Biomedical and Biotechnology Research Journal (BBRJ)*, *3*(3), 196. https://doi.org/10.4103/bbrj.bbrj_82_19

Ochora, D. O., Kakudidi, E., Namukobe, J., Heydenreich, M., Coghi, P., Yang, L. J., Mwakio, E. W., Andagalu, B., Roth, A., Akala, H. M., Wong, V. K. W., & Yenesew, A. (2022). A new benzophenone, and the antiplasmodial activities of the constituents of Securidaca longipedunculata Fresen (Polygalaceae). *Natural Product Research*, *36*(11), 2758–2766. https://doi.org/10.1080/14786419.2021.1925272

Ochora, D. O., Kakudidi, E. K., Namukobe, J., Ipulet, P., Wakoli, D. M., Okore, W., Mwakio, E. W., Yeda, R. A., Cheruiyot, A. C., Juma, D. W., Andagalu, B., Roth, A. L., Ogutu, B. R., Yenesew, A., & Akala, H. M. (2022). Synergism in antiplasmodial activities of artemether and lumefantrine in combination with Securidaca longipedunculata Fresen (Polygalaceae). *Plants*, *11*(1), Article 1. https://doi.org/10.3390/plants11010047

Ojewole, J. A. O. (2008). Analgesic, anti-inflammatory and hypoglycaemic effects of Securidaca longepedunculata (Fresen) [Polygalaceae] root-bark aqueous extract. *Inflammopharmacology*, *16*(4), 174–181. https://doi.org/10.1007/s10787-007-0016-7

Ouedraogo, L., Endl, J., Sombié, P. A. E. D., Schaefer, H., & Kiendrebeogo, M. (2020). Ethnobotanical use and conservation assessment of medicinal plants sold in markets of Burkina Faso. *Ethnobotany Research and Applications*, *20*, 1–25.

Ouédraogo, P., Bationo, B. A., Sanou, J., Traoré, S., Barry, S., Dayamba, S. D., Bayala, J., Ouédraogo, M., Soeters, S., & Thiombiano, A. (2017). Uses and vulnerability of ligneous species exploited by local population of northern Burkina Faso in their adaptation strategies to changing environments. *Agriculture & Food Security*, *6*(1), 15. https://doi.org/10.1186/s40066-017-0090-z

Plants of the World Online, POWO (2025). Distribution of Securidaca longepedunculata source. https://powo.science.kew.org/taxon/urn:lsid:ipni.org:names:692714-1#synonyms (accessed 17 September 2025).

Semenya, S. S., Potgieter, M. J., & Erasmus, L. J. C. (2013). Indigenous plant species used by Bapedi healers to treat sexually transmitted infections: Their distribution, harvesting, conservation and threats. *South African Journal of Botany*, *87*, 66–75. https://doi.org/10.1016/j.sajb.2013.03.001

Stevenson, P. C., Arnold, S. E. J., & Belmain, S. R. (2014). Pesticidal plants for stored product pests on small-holder farms in Africa. In D. Singh (Ed.), *Advances in Plant Biopesticides* (pp. 149–172). Springer India. https://doi.org/10.1007/978-81-322-2006-0_9

Suzana Augustino. (2011). Medicinal resources of the Miombo woodlands of Urumwa, Tanzania: Plants and its uses. *Journal of Medicinal Plants Research*, *5*(27). https://doi.org/10.5897/jmpr10.517

Tabuti, J. R. S., Muwanika, V. B., Arinaitwe, M. Z., & Ticktin, T. (2011). Conservation of priority woody species on farmlands: A case study from Nawaikoke sub-county, Uganda. *Applied Geography*, *31*(2), 456–462. https://doi.org/10.1016/j.apgeog.2010.10.006

Tevfik, A. S., & Yegorova, N. A. (2020). Clonal micropropagation of Thymus vulgaris L. *E3S Web of Conferences*, *224*, 04001. https://doi.org/10.1051/e3sconf/202022404001

Tewelde, F., & Mesfin, M. (2020). Ethnobotanical use and conservation of plants biodiversity by the local community of Welkait Wereda, Western Tigray, Ethiopia. *Advances in Life Science and Technology*, *83*, 1–13. https://doi.org/10.7176/ALST/83-01

Tiawoun, M. A. P., Ligavha-Mbelengwa, M., & Tshisikhawe, M. P. (2015). Reproductive biology towards the conservation of Securidaca longipedunculata Fresen in the Nylsvley Nature Reserve, Limpopo Province, South Africa. *South African Journal of Botany*, *98*, 202–203. https://doi.org/10.1016/j.sajb.2015.03.127

Tiawoun, M. A. P., Tshisikhawe, M. P., & Ligavha-Mbelengwa, M. (2017). Seed germination and seedling establishment of Securidaca longepedunculata (Polygalaceae). *Pakistan Journal of Botany*, *49*, 1009–1016.

Tikisa, T., Abdissa, D., & Abdissa, N. (2019). Chemical constituents of Securidaca longipedunculata root bark and evaluation of their antibacterial activities. *Ethiopian Journal of Education and Sciences*, *14*(2), 1–8. Retrieved from https://journals.ju.edu.et/index.php/ejes/article/view/665

Tsay, H. S., Gau, T. G., & Chen, C. C. (1989). Rapid clonal propagation of Pinellia ternata by tissue culture. *Plant Cell Reports*, *8*(8), 450–454. https://doi.org/10.1007/BF00269046

Tshisikhawe, M. P., Baloyi, O., Ligavha-Mbelengwa, M. H., & Bhat, R. B. (2012). The population ecology of *Securidaca longepedunculata* Fresen. *In the Nylsvley Nature Reserve, Limpopo Province, South Africa*. *81*(1), 107–112.

van Andel, T., Myren, B., & van Onselen, S. (2012). Ghana's herbal market. *Journal of Ethnopharmacology*, *140*(2), 368–378. https://doi.org/10.1016/j.jep.2012.01.028

van Andel, T. R., Croft, S., van Loon, E. E., Quiroz, D., Towns, A. M., & Raes, N. (2015). Prioritizing West African medicinal plants for conservation and sustainable extraction studies based on market surveys and species distribution models. *Biological Conservation*, *181*, 173–181. https://doi.org/10.1016/j.biocon.2014.11.015

Vidyagina, E. O., Kharchenko, N. N., & Shestibratov, K. A. (2021). Efficient cryopreservation of Populus tremula by in vitro-grown axillary buds and genetic stability of recovered plants. *Plants*, *10*(1), Article 1. https://doi.org/10.3390/plants10010077

Wang, Y. R., Hanson, J., & Mariam, Y. W. (2007). Effect of sulfuric acid pretreatment on breaking hard seed dormancy in diverse accessions of five wild Vigna species. *Seed Science and Technology*, *35*(3), 550–559. https://doi.org/10.15258/sst.2007.35.3.03

World Health Organization (2022). African Traditional Medicine Day 2022. *WHO | Regional Office for Africa* (accessed: 17 September 2025). www.afro.who.int/regional-director/speeches-messages/african-traditional-medicine-day-2022

Wyk, B. V., & Gericke, N. (2000). People's plants: *A guide useful plants for southern Africa* (pp. 351–pp).

Zéphirin, O. H., Guidawa, F., Herbert, A., Youhana, D., Bruno, N. N. J., & Marie, M. P. (2019). Vegetative propagation by root segment cuttings of Securidaca longepedunculata Fresen: Effect of diameter and insertion mode of root segment cuttings in the substrate. *Research Journal of Agriculture and Forestry Sciences, ISSN*, *2320*, 6063.

Zulu, D., Thokozani, B., Sileshi, G., Teklehaimanot, Z., Gondwe, D., Sarasan, V., & Stevenson, P. (2011, July 3). Propagation of the African medicinal and pesticidal plant, Securidaca longepedunculata. *African Journal of Biotechnology*, *10*(32), 5988–5992. https://doi.org/10.5897/AJB10.2603

Chapter 10

Dermatological potentials of *Securidaca* spp. and their phytochemical properties

Abdul Awal Chowdhury Masud,[1] Nahid Akter,[2] Daniel Jesuwenu Ajose,[3] Tabassum Tamima,[4] Md Saidur Rahman,[5] Md Shamsuzzaman,[6] Khairul Kabir,[7] and Delower Hossain[8]*

[1]Department of Agricultural and Environmental Sciences – Production, Landscape, Agroenergy (DiSAA), Università degli Studi di Milano (UNIMI), Milano, Italy
Department of Agronomy, Faculty of Agriculture, Sher-e-Bangla Agricultural University (SAU), Dhaka, Bangladesh
[2]Department of Plant Pathology, Bioregulation Science, United Graduate School of Agriculture, Tokyo University of Agriculture and Technology, Tokyo, Japan
Department of Plant Pathology, Faculty of Agriculture, Sher-e-Bangla Agricultural University (SAU), Dhaka, Bangladesh
[3]Antimicrobial Resistance and Phage Bio-Control Research Lab, Department of Microbiology, School of Biological Sciences, Faculty of Natural and Agricultural Sciences, North-West University, Mmabatho, South Africa
School of Biology and Environmental Sciences, Faculty of Agriculture and Natural Sciences, University of Mpumalanga, Mpumalanga, South Africa
[4]Department of Biology, St. Cloud State University, MN, USA
[5]Research Institute for Northern Agriculture, Faculty of Science and Technology, Charles Darwin University, NT, Australia
Department of Agricultural Botany, Faculty of Agriculture, Sher-e-Bangla Agricultural University (SAU), Dhaka, Bangladesh
[6]Faculty of Science, School of Life and Environmental Sciences, The University of Sydney
Department of Biochemistry, Faculty of Agriculture, Sher-e-Bangla Agricultural University (SAU), Dhaka, Bangladesh
[7]Department of Horticulture, Faculty of Agriculture, Sher-e-Bangla Agricultural University (SAU), Dhaka, Bangladesh
[8]Department of Veterinary Medicine and Animal Sciences (DIVAS), Università degli Studi di Milano (UNIMI), Lodi, Italy
Department of Medicine and Public Health, Faculty of Animal Science and Veterinary Medicine, Sher-e-Bangla Agricultural University (SAU), Dhaka, Bangladesh

*Correspondence: chy.masud3844@sau.edu.bd, abdul.masud@unimi.it

 DOI: 10.1201/9781003544463-10

10.1 INTRODUCTION

The plant kingdom is a vital source of natural products used across industries, such as pharmaceuticals, agriculture, food, and cosmetics. These botanical constituents offer therapeutic benefits, eco-friendly solutions as pesticides and growth enhancers, and play essential roles in food preservation, flavoring, and traditional and folk medicinal practices (Jahan et al. 2019). In recent years, interest in plant-derived compounds has increased due to their versatile applications and the growing consumer demand for natural, organic, and eco-friendly products (Shemishere et al. 2020, Hernandez et al. 2021). Approximately 25% of modern pharmaceutical products in today's pharmacopeia are plant-derived, with many others being synthetic versions inspired by plant-based compounds (Kadiri et al. 2013). Traditional health practitioners in developing countries have relied on medicinal plants as a primary form of healthcare, with approximately 80% of the population from developing countries depending on these traditional remedies (Poudiougou et al. 2021). This large dependency is due to inadequate access to modern healthcare infrastructure and limited access to conventional medications. In addition to their therapeutic efficacy, these botanical constituents hold significant cultural importance and are now increasingly integrated into modern dermatological formulations to enhance skin health and wellness (Curran 2017).

Among the diverse flora utilized in ethnomedicine, the genus *Securidaca* (family: Polygalaceae) stands out for its extensive pharmacological potential, including anti-inflammatory, antimicrobial, and antioxidant effects (Pang et al. 2021). This genus comprises woody climbers, shrubs, and small trees that are renowned in traditional systems for treating a wide array of ailments, ranging from inflammation and pain to infectious diseases (Pang et al. 2021; Obasi et al. 2018). The genus *Securidaca* includes approximately 80 recognized species, and its members share a rich phytochemical profile, including flavonoids, xanthones, alkaloids, and saponins, which underpins their biological activities (Makangara et al. 2024; Afrisham et al. 2015). The potential of *Securidaca* extracts in promoting wound healing and enhancing skin hydration further expands their therapeutic scope. Notably, some species, such as *S. longipedunculata, S. diversifolia,* and *S. inappendiculata,* are becoming popular for their dermatological benefits, including their ability to promote wound healing, alleviate skin inflammation, and soothe skin irritation (Ndamitso et al. 2013). In addition to their therapeutic applications, the *Securidaca* genus also has potential for addressing emerging challenges such as antimicrobial resistance. Recent studies have demonstrated that certain flavonoids derived from *Securidaca* extracts possess significant antibacterial properties against resistant strains of bacteria and fungi, including *Staphylococcus aureus* and *Candida albicans* (Moiketsi et al. 2023, Sani et al. 2023).

The *Securidaca* genus has a pantropical distribution, thriving primarily in the tropical and subtropical regions of Africa, Asia, and the Americas. Most of the recognized species are native to the Americas, whereas fewer species are recorded in Africa and the Indo-Malaysian region (Johnson 1987). Despite its more limited species diversity in Africa, the genus has immense economic and medicinal value. *S. longipedunculata*, commonly known as the "violet tree" or "Rhodesian violet," is widespread across the hot, arid, and humid climates of sub-Saharan Africa, ranging from Guinea and Nigeria in the west to Angola and South Africa in the south (Gaobotse et al. 2023; Mongalo et al. 2015). Data from the Global Biodiversity Information Facility (GBIF) document over 2,700 occurrences, confirming its resilience across diverse soil types and moisture levels in savannas and woodlands (GBIF 2024; Zulu et al. 2011).

Dermatological conditions, including wounds, skin infections, and premature aging caused by oxidative stress, represent a significant burden on global health. The *Securidaca* genus is traditionally used in treating skin inflammation, soothing irritation, and promoting wound healing due to its pharmacological properties (Ndamitso et al. 2013). Species including *S. longipedunculata*, *S. diversifolia*, and *S. inappendiculata* demonstrate particular therapeutic promise against such different dermatological conditions (Guo and Jee 2021). Notably, recent research highlights that *Securidaca*-derived bioactive compounds inhibit multidrug-resistant bacteria and fungi, which are common in skin infections (Moiketsi et al. 2023).

This chapter provides a comprehensive overview of the dermatological potential of the *Securidaca* genus. It explores the botanical features and traditional uses of these plants, detailed phytochemical profiles relevant to skin health, and the pharmacological mechanisms driving their wound healing and antiaging effects. Furthermore, it addresses the challenges and opportunities in translating these traditional remedies into modern, safe, and effective skin care formulations.

10.2 BOTANICAL AND ETHNOMEDICINAL USES

10.2.1 Botanical description

The genus *Securidaca* belongs to the family Polygalaceae and comprises approximately 80 species of woody climbers, shrubs, and small trees (Obasi et al. 2018). The height and branching patterns can vary significantly among species, but some species can grow up to 12 m long (Keshebo and Choudhury 2015). While the genus displays significant morphological diversity, its members generally feature simple, alternate leaves that vary in shape from linear to ovate or lanceolate (Aymard-Corredor 2022). A distinguishing characteristic of the genus is its flowers, which are typically arranged in racemes or panicles and exhibit irregular shapes such as pea flowers. These flowers are often vibrant, appear in shades of purple, pink, or white, and possess five petals (Suleiman et al. 2022). The fruit is a dehiscent capsule, often a winged samara adapted for wind dispersal (Johnson 1987).

S. longipedunculata Fresen, widely known as the "violet tree" or "Rhodesian violet," is the most prominent species in Africa and serves as a representative example of the genus (Gaobotse et al. 2023). In Nigeria, owing to its high cultural significance across different ethnic groups, the plant is known by various names. It is called "Uwar magunguna or Sanya" in Hausa, "Ezeogwu" in Igbo, "Ipeta" in Yoruba, and "Aaalali" in the Fulani-speaking community (Namadina et al. 2020). This species has been the focus of numerous scientific studies, highlighting its ecological, cultural, and pharmacological importance (Zulu et al. 2011, Barrera-Guzmán et al. 2021, Lawal et al. 2012). It grows as a small tree or shrub up to 9 m in height, with pale gray, smooth bark. Its sweet-scented flowers bloom in early summer, displaying colors ranging from pink to violet-purple (Suleiman et al. 2022), whereas its fruits are oblong, purplish-green, and heavily veined. Owing to its medicinal value, the stem bark and roots of *S. longipedunculata* are among the most heavily traded plant materials in African traditional medicine markets (Tabuti et al. 2012).

10.2.2 Ethnomedicinal applications and traditional formulations

Ethnomedicinal practices across Africa (Mongalo et al. 2015), Asia (Qian et al. 2020), and South America (Suleiman et al. 2022) extensively utilize *Securidaca* species to treat a spectrum of ailments, with a particular emphasis on dermatological and inflammatory conditions (Guo and Jee 2021). The root is the most frequently utilized plant part and is often valued for its potent analgesic and anti-inflammatory properties. In African traditional medicine, *S. longipedunculata* is called "Uwar magunguna" in the Hausa language, which means "The mother of all medicines". It is employed to treat skin infections, wounds, snakebites, and sexually transmitted infections (STIs), as well as systemic conditions such as malaria, rheumatism, and epilepsy (Mongalo et al. 2015). Various parts of the *S. longipedunculata* plant, such as the roots, leaves, and bark, alone or in combination with other plant parts, are mixed to make unique formulations that are used to treat ailments such as pain, inflammation, fever, respiratory issues, and digestive disorders (Figure 10.1). In Asia, specifically southern China and Vietnam, *S. inappendiculata* is traditionally used to manage rheumatoid arthritis and fractures, likely because of its strong anti-inflammatory and antioxidative effects (Qian et al. 2020). Traditional formulations are diverse and often involve synergistic combinations with other plants. For dermatological issues, roots or leaves are commonly crushed into pastes or poultices and applied topically to wounds or inflamed skin. Systemic ailments are

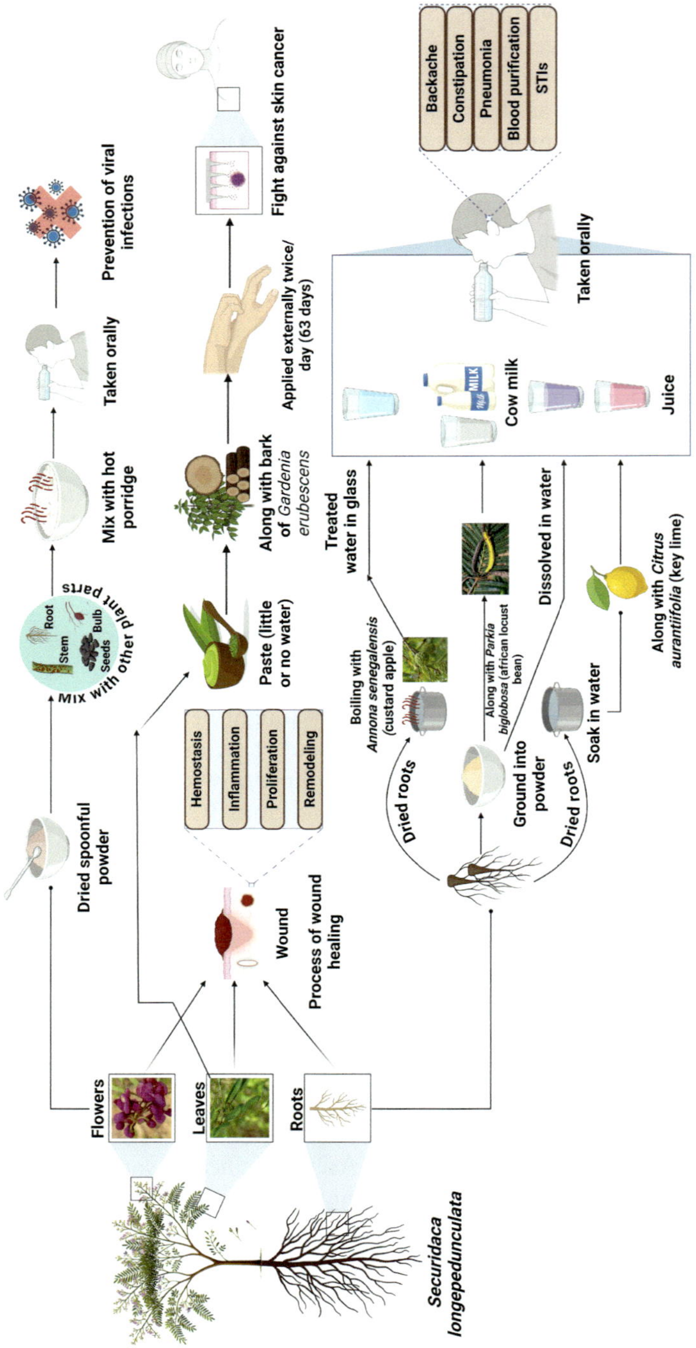

Figure 10.1 Traditional therapeutic formulations using different parts of *S. longipedunculata*.

typically treated with decoctions or infusions consumed orally. The specific ethnobotanical uses and their corresponding traditional formulations are detailed in Table 10.1.

10.3 PHYTOCHEMISTRY: THE BIOACTIVE PROFILE

The therapeutic efficacy of the *Securidaca* genus is attributed to its diverse array of secondary metabolites (Mongalo et al. 2015). Phytochemical screening of extracts from *S. diversifolia*, *S. longipedunculata*, and *S. inappendiculata* revealed a rich composition of xanthones, flavonoids, alkaloids, saponins, triterpenes, and phenolic acids (Table 10.2). The specific distribution of these compounds varies by plant part; for example, root bark is particularly rich in methyl salicylate and xanthones, whereas leaves often contain lower concentrations of certain metabolites than roots do (Jayasekara et al. 2002; Musa et al. 2013). Table 10.2 presents a comprehensive overview of the major phytochemical compounds isolated from *Securidaca* spp., listing their chemical class, key compounds, structural formulas, and traditional and potential therapeutic applications and herbal uses.

10.4 PHARMACOLOGICAL ACTIVITIES

The bioactive compounds identified in the *Securidaca* genus exhibit a broad spectrum of pharmacological activities. These properties are particularly relevant to dermatology, offering potential therapeutic avenues for skin infections, wound repair, inflammation, and age-related skin deterioration. Several studies have confirmed that *S. longipedunculata* has broad pharmacological potential and is a promising natural alternative to synthetic drugs for addressing multidrug resistance. All the plant parts (roots, stems, bark, leaves, and flowers) contain antimicrobial compounds rich in *n*-hexane, ethyl acetate, chloroform, and methanol extracts. These extracts yield biologically active secondary metabolites responsible for plant antimicrobial, anti-inflammatory, analgesic, antioxidant, and antifungal properties against various pathogens, including bacteria, fungi, and parasites (Hassan et al. 2024; Kporvie 2024), as presented in Table 10.3. Extracts from *S. longipedunculata* and *S. inappendiculata* have demonstrated significant efficacy against a wide range of bacterial species that cause different dermatological and systemic infections, such as *Streptococcus* spp., *Staphylococcus* spp., *Escherichia coli*, *Serratia* spp., *Enterococcus* spp., *Shigella* spp., *Salmonella* spp., *Pseudomonas* spp., *Klebsiella* spp., *Bacillus subtilis*, and *Mycobacterium* spp. (Hassan et al. 2024; Namadina et al. 2020; Tikisa et al. 2019; Ferreira et al. 2022). Moreover, this genus is also effective against different fungal infections caused by *Candida* spp., *Aspergillus* spp., and *Microsporum* spp. (Hassan et al. 2024; Sani et al. 2023). In addition, they

Table 10.1 Ethnomedicinal uses and traditional formulations of *Securidaca* spp.

Sl. No.	*Species*	*Country*	*Plant Parts*	*Traditional Uses/Diseases*	*References*
1	*S. diversifolia*	Argentina	Leaves	Diabetes	Ezuruike and Prieto 2014
2	*S. inappendiculata*	China	Roots, stems, leaves, stem bark	Diabetes, wounds, fractures, pain relief, rheumatoid arthritis, infections, inflammation	Zuo et al. 2014 Pang et al. 2021
		Vietnam	Roots	Anti-inflammatory and pain relief	Nguyen et al. 2020
3	*S. longipedunculata*	Angola	Roots	Pregnancy problems, scabies, tuberculosis, epilepsy, rheumatism, hepatitis, madness, snakebite, malaria, cancer	Ngulde et al. 2019 Mongalo et al. 2015
		Botswana	Roots	Coughs, aphrodisiac, tuberculosis	Motlhanka and Nthoiwa 2012
		Burkina Faso	Stem bark, leaves, roots	Skin diseases, constipation, malaria, vomiting, sedative, wounds, neurological problems, poisoning, musculoskeletal issues, malaria	Ouedraogo et al. 2020
		Brazil	Leaves, roots	Rheumatism, malaria, diabetes, arthritis	Saraiva et al. 2015; Santos-Buelga et al. 2023
		Burundi	Roots	Laxative, purgative, prevents abortion, intestinal parasites, psychosis	Desta 1993
		Congo	Roots, leaves	Wound healing	Valentin et al. 2024

Sl. No.	*Species*	*Country*	*Plant Parts*	*Traditional Uses/Diseases*	*References*
		Ethiopia	Roots	Cough, muscular problems, herpes, malaria	Awoke et al. 2024
		Guinea Conakry	Roots	Antiseptic, anti-inflammatory	Magassouba et al. 2007
		Kenya	Roots and leaves, whole plant	Malaria fever, coughs, pain relief, antiaging	Wanzala et al. 2016 Motlhanka and Nthoiwa 2012
		Mali	Roots	Dysmenorrhea, headache, anti-suicidal, menstruation pain	Nordeng et al. 2013
		Nigeria	Roots, leaves, stem bark	Anti-abortion, constipation, coughs, fever, pneumonia, sexual boost, toothache, tuberculosis, rheumatism, dislocated jaw, headaches, skin cancer, skin infections	Mustapha 2013
		South Africa	Roots	Flu, blood purifier, aphrodisiac, psychoactive purposes, contraceptive, malaria	Semenya and Potgieter 2013 Iyamah and Idu 2015
		Tanzania	Roots	Meningitis, HIV	Kisangau et al. 2007
		Togo	Leaves	Liver diseases, malaria, hemorrhoids, trypanosomiasis, headache, snakebites, tuberculosis, syphilis	Kpodar et al. 2016
		Uganda	Roots	Fever, malaria, ascariasis, diarrhea, headache, body pain, epilepsy, flu, hernia, madness	Tabuti et al. 2012 Hamill et al. 2003
		Zimbabwe	Roots	Venereal diseases, syphilis, pains, fever, epilepsy, pneumonia, tuberculosis, epilepsy, pain, STIs	Maroyi 2013

Table 10.2 Major phytochemical compounds isolated from *Securidaca* spp. and their potential uses

Sl. No.	*Chemical Class*	*Species*	*Key Compounds*	*Structural Formula*	*Potential Herbal Uses*	*References*
1	Alkaloids	*S. longipedunculata*	5-O-prenyl-1-hydroxy-2,3, 6, 7,8-pentamethoxyxanthone (Securinine) ($C_{24}H_{28}O_8$)	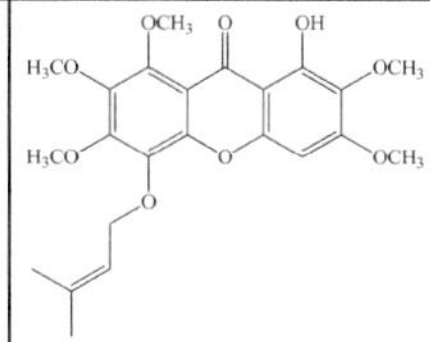	• Neuroprotective • Anti-inflammatory effects • Reduce inflammation in acne • Antimicrobial	Klochkov and Neganova 2021 Sani et al. 2023 Mongalo et al. 2015
		S. longipedunculata	Securidine ($C_{13}H_{15}NO_2$)	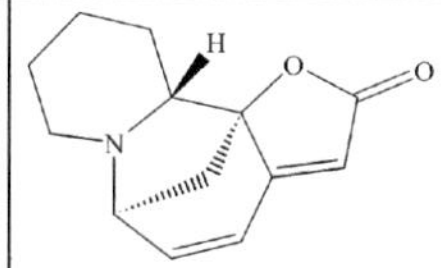	• Treating neurological disorders • Neurostimulatory • Cognitive health support	Tian et al. 2018
2	Benzoates and esters	*S. longipedunculata*	Benzyl-2-hydroxy-6-methoxybenzoate ($C_{15}H_{14}O_4$)	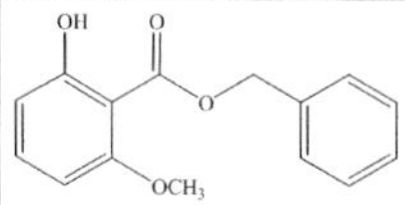	• Herbal remedies • Cosmetic applications • Antioxidant • Anti-inflammatory	Lognay et al. 2000
		S. longipedunculata	Benzyl-3-hydroxy-2-methoxybenzoate (R1 = H, R2 = OOH)	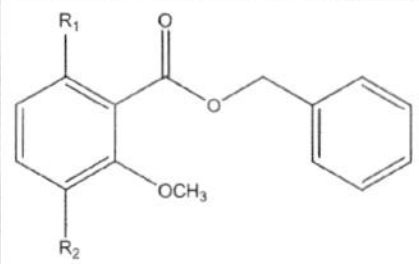	• Herbal remedies • Cosmetic applications • Antioxidant • Anti-inflammatory	Al-Snafi 2020s

Sl. No.	*Chemical Class*	*Species*	*Key Compounds*	*Structural Formula*	*Potential Herbal Uses*	*References*
		S. inappendiculata	Coumarins ($C_9H_6O_2$)	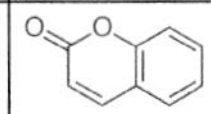	Fragrant compounds with • Anti-inflammatory, antimicrobial effects • Promote skin tone • Reduce hyperpigmentation • Anticoagulant property	Datagni et al. 2021
3	Fatty acids	*S. longipedunculata* *S. inappendiculata*	Linoleic acid ($C_{18}H_{32}O_2$)		• Maintain skin barrier function • Prevent transepidermal water loss • Reduce inflammation • Skin repair	Jiang et al. 2018
4	Flavonoids	*S. longipedunculata* *S. diversifolia*	Quercetin ($C_{15}H_{10}O_7$)	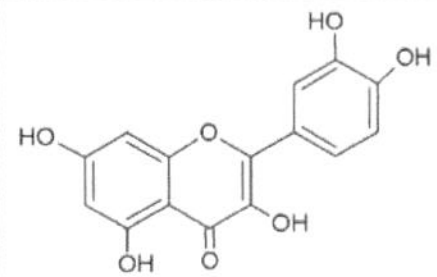	• Neutralize free radicals • Accelerate slow skin aging • Reduce skin redness, irritation	Jiang et al. 2018
5	Phenolic compounds	*S. longipedunculata*	Sinapic acid ($C_9H_{10}O_4$)	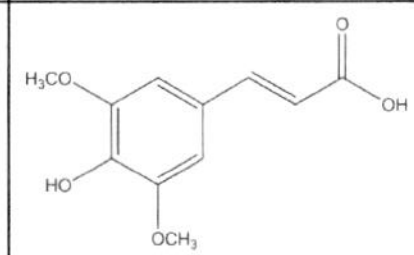	• Anti-inflammatory • Prevent premature aging • Neuroprotective effects • Reduce inflammation in arthritis	Lee et al. 2021 Wang et al. 2021
		S. longipedunculata	Methyl-2-hydroxy-6-methoxybenzoate ($C_9H_{10}O_4$)	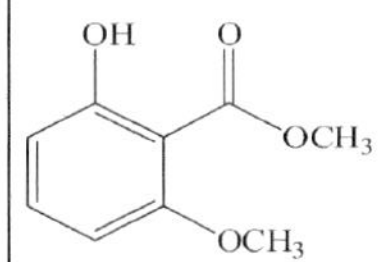	• Anti-inflammatory • Neuroprotective effects • Reduce inflammation in arthritis • Support skin rejuvenation	Karama et al. 2022

Table 10.2 (Continued)

Sl. No.	*Chemical Class*	*Species*	*Key Compounds*	*Structural Formula*	*Potential Herbal Uses*	*References*
		S. longipedunculata *S. inappendiculata*	Caffeic acid ($C_9H_8O_4$)	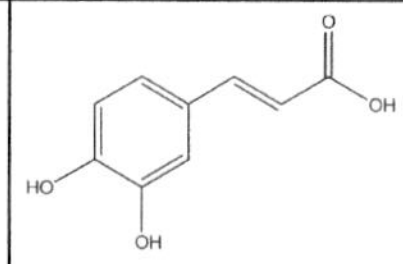	• Antioxidant • Reduce inflammation • Reduce oxidative stress • Reduce risk of chronic diseases	Alam et al. 2022
		S. inappendiculata *S. longipedunculata*	p-Coumaric acid ($C_9H_8O_3$)	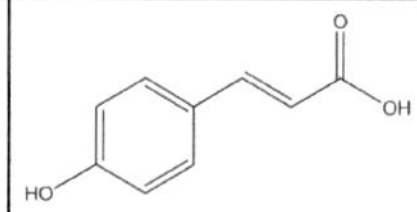	• Antimicrobial • Prevent infections • Inhibit growth of bacteria, fungi • Protect from UV radiation	Lou et al. 2012 Wu et al. 2016 Li et al. 2022
		S. longipedunculata *S. inappendiculata*	Cinnamic acid ($C_9H_8O_2$)	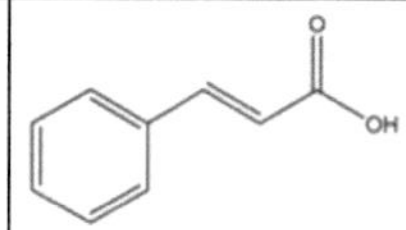	• Alleviate respiratory infections • Reduce mouth bacterial growth • Antimicrobial • Anti-inflammatory • Oral health support	Manogaran et al. 2023 Carrol et al. 2020 Dalir Abdolahinia et al. 2023
6	Saponins	*S. longipedunculata* *S. inappendiculata*	Aescin ($C_{57}H_{92}O_{28}$)	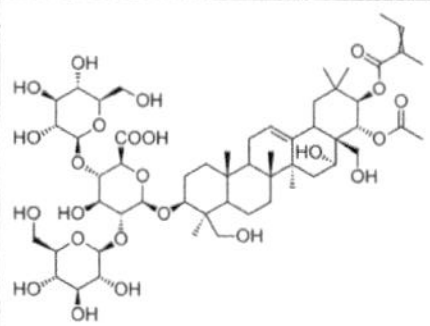	• Skin cleansing • Moisturizing • Wound heazling • Anti-inflammatory • Antimicrobial properties	Tian et al. 2018

Sl. No.	*Chemical Class*	*Species*	*Key Compounds*	*Structural Formula*	*Potential Herbal Uses*	*References*
	Sterols and related compounds	*S. longipedunculata*	Methyl salicylate ($C_8H_8O_3$)	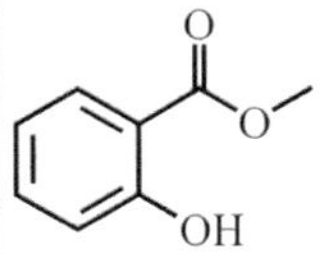	• Pain relief • Anti-inflammatory agent • Remove muscle pain and arthritis	Jayasekara et al. 2002
7		*S. longipedunculata* *S. inappendiculata*	Presenegenin ($C_{13}H_{14}O_3$)	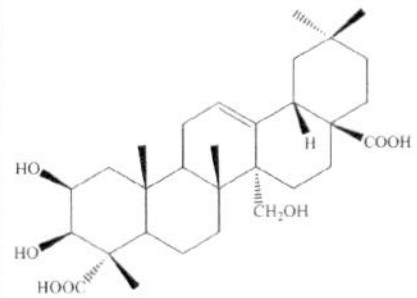	• Making herbal formulations • Improve cognitive health • Neuroprotective • Anti-inflammatory effects	Pandita and Pandita 2021
		S. inappendiculata *S. longipedunculata* *S. diversifolia*	β-Sitosterol ($C_{27}H_{49}O$)	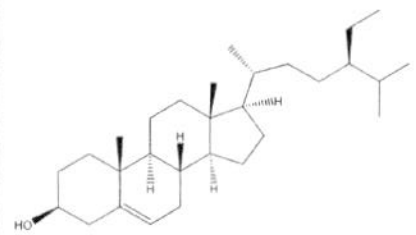	• Herbal supplements • Alleviate prostatic hyperplasia • Lowers cholesterol • Supports BPH	Sharma et al. 2017 Mădălina et al. 2024
8	Triterpenoids	*S. inappendiculata* *S. longipedunculata*	Oleanolic acid ($C_{30}H_{48}O_3$)	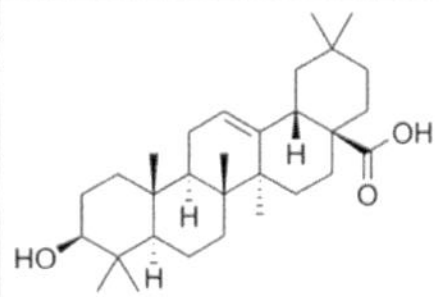	• Skin barrier repair • Immune function • Reduce inflammation • Promote skin healing	Namadina et al. 2020 Karama et al. 2022
9	Xanthones	*S. longipedunculata*	4-Methoxy-benzo [1, 3] dioxol-5-yl-phenylmethanone, 1,7-Dihydroxy-4-methoxyxanthone ($C_{15}H_{12}O_4$)	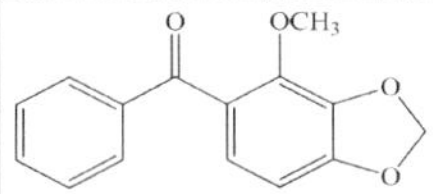	• Antioxidant • Potential anticancer effects	Melaku et al. 2022 Wang et al. 2017

(continued)

Table 10.2 (Continued)

Sl. No.	*Chemical Class*	*Species*	*Key Compounds*	*Structural Formula*	*Potential Herbal Uses*	*References*
		S. inappendiculata	1,6,8-Trihydroxy-2,3,4,7-tetramethoxyxanthone ($C_{19}H_{18}O_{10}$)	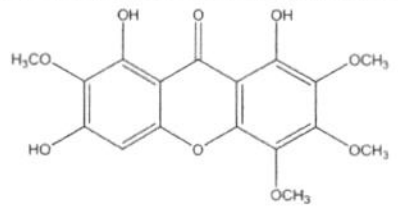	• Antioxidant • Anti-inflammatory effects	Melaku et al. 2022 Wang et al. 2017 Gunter et al. 2020
		S. longipedunculata	1,7-Dihydroxy-4-methoxyxanthone ($C_{19}H_{18}O_{10}$)	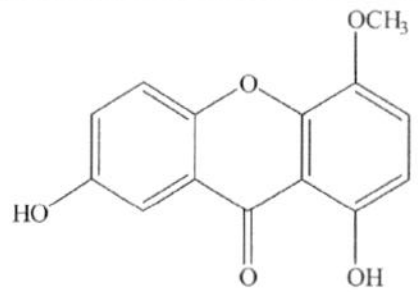	• Antioxidant • Anti-inflammatory effects	Mohan et al. 2018
		S. longipedunculata *S. inappendiculata*	2-Methoxy-3,4-methylened-ioxybenzophenone ($C_{15}H_{12}O_4$)	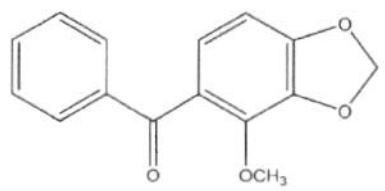	• Use as herbal medicine • Use as herbal cosmetics • Antioxidant • Anti-inflammatory	Wong et al. 2023s

Table 10.3 Antibacterial, antifungal, and antiparasitic and antiprotozoal activities of different parts of *Securidaca longipedunculata*

Sl. No.	*Activity*	*Plant Parts*	*Solvent (Extract)*	*Inhibition Zone (mm)/ MIC/(mg/mL)*	*Microbes Inhibited*	*References*
1	Antibacterial activity	Root	Methanol	Inhibition zone: 28 mm	*Streptococcus pyogenes* and *Staphylococcus aureus*	Sani et al. 2023
			70% Methanol	MIC: of 0.45 mg/mL and 0.23 mg/mL	*Serratia marcescens* and *Shigella flexneri*	Karou et al. 2012
			Chloroform	MIC: 0.591 mg/mL	*Salmonella typhi* and *Pseudomonas aeruginosa*	Ndamitso et al. 2013 Tikisa et al. 2019
			Aqueous	MIC: of 6.25 mg/mL	*Salmonella typhi*	Namadina et al. 2020
		Leaves	Acetone	MIC: 40.1 mg/mL	*Mycobacterium tuberculosis*	Sani et al. 2023
			Methanol, Ethyl acetate	MIC: 0.0625–0.25 mg/mL Where Bactericides/ fungicides MIC: 0.25–0.5 mg/mL	*Enterococci, Enterococcus faecalis, Staphylococcus aureus, Klebsiella pneumoniae, Salmonella typhimurium, Pseudomonas fluorescens*, and *Escherichia coli*	Hassan et al. 2024 Tikisa et al. 2019 Namadina et al. 2020
		Root bark	Chloroform	Inhibition zone: 28 mm	*Staphylococcus aureus*	Musa et al. 2013
			n-Hexane, Ethyl acetate	Inhibition zone: 14–19 mm	*Streptococcus pyogenes, Pseudomonas fluorescens*, and *Klebsiella pneumoniae*	Toudji et al. 2018 Datagni et al. 2021
			Chloroform	Inhibition zone: 19–21 mm	*Escherichia coli, Salmonella typhi, Shigella dysenteriae, Klebsiella pneumonieae*, and *Staphylococcus. aureus*,	Abubakar et al. 2018
		Light stem and root bark	Distilled Water and Methanol	MIC: 0.63 mg/ml	*Escherichia coli, Pseudomonas aeruginosa*, and *Enterococcus faecalis*	Shai et al. 2024
		Root	Acetone	MIC of 0.02 mg/mL	*Bacillus subtilis* and *Salmonella typhi*	Tikisa et al. 2019

(continued)

Table 10.3 (Continued)

Sl. No.	*Activity*	*Plant Parts*	*Solvent (Extract)*	*Inhibition Zone (mm)/ MIC/(mg/mL)*	*Microbes Inhibited*	*References*
			Acetone	MIC of 0.10 mg/mL	*Escherichia coli* and *Pseudomonas aeruginosa*	Tikisa et al. 2019
			n-Hexane	MIC: 0.0312–40.250 mg/mL	*Mycobacterium* spp., *M. smegmatis*, *M. bovis*, *M. tuberculosis*, and *M. avium*	Wang et al. 2021 Ferreira et al. 2022
		Plant	Essential oil	MIC of 12.79 mg/mL	*Escherichia coli*	Hassan et al.2024
2	Antifungal activity	Leaf	Methanol, Ethyl Acetate	MIC: 0.0625–0.25 mg/mL Where Bactericides MIC: 0.25–0.5 mg/mL	*Candida albicans*, *Candida krusei*, *Aspergillus niger*, *Aspergillus fumigatus*, and *Microsporum canis*	Hassan et al. 2024 Namadina et al. 2020 Datagni et al. 2021
		Leaf	Methanol	28 mm	*Candida albicans*	Sani et al. 2023
3	Antiparasitic and antiprotozoal activity		Methanol	Inhibited motility	*Trypanosoma brucei brucei* and *Trypanosoma congolense*	Adiele et al. 2013
		Root bark	Aqueous and Methanol	Antitrypanosomal activity	*Trypanosoma brucei rhodesiense*	Imam et al. 2017
				Larvicidal effect	*Heligmosomoides contortus* and *Heligmosomoides polygyrus*	Adiele et al. 2013
		Root	Aquous	Number of viable parasites	*Trichomonas vaginalis*	Imam et al. 2017
		Stem bark	Petroleum Ether	Inhibited motility	*Trypanosoma brucei*	Mongalo et al. 2015

Abbreviation: MIC: Minimum inhibitory concentration.

also showed efficiency against different parasites and protozoa, including *Heligmosomoides* spp., *Trypanosoma* spp., and *Trichomonas* spp. (Adiele et al. 2013; Imam et al. 2017).

10.4.1 Anti-inflammatory and analgesic effects

Inflammation is a central component of many skin disorders, including eczema and dermatitis. *Securidaca* spp. possess potent anti-inflammatory properties, comparable in some studies to nonsteroidal anti-inflammatory drugs (NSAIDs) (Ahmed et al. 2024; Nair et al. 2024). *S. longipedunculata* serves as a promising natural alternative to NSAIDs for managing inflammation. Its anti-inflammatory activity is largely attributed to methyl salicylate, which constitutes 90% of root bark volatiles and blocks prostaglandin and histamine production (Jayasekara et al. 2002). Phenolic acids and resins also contribute to this effect, showing a stronger positive correlation with anti-inflammatory activity than flavonoids do (Lema et al. 2024). The bark extracts exhibit immunomodulatory potential, with aqueous extracts stimulating IL-2 and root bark extracts dramatically increasing anti-inflammatory IL-10 production by 500-fold (Nair et al. 2024). Furthermore, a 50% aqueous extract of the bark of *S. longipedunculata* also inhibited nitric oxide production in stimulated macrophages by 51.3% according to 2,2'-azino-bis(3-ethylbenzothiazoline-6-sulfonic) (ABTS), and 1,1-Diphenyl-2-picryl hydrazyl radical (DPPH) tests (Muanda et al. 2010). In inflammation models, its petroleum ether (65.63%), methanol (53.13%), and whole methanol extracts (40.63%) significantly reduce inflammation (Okoli et al. 2005). Hydroalcoholic extracts have demonstrated membrane-stabilizing effects that exceed those of diclofenac, and leaf extracts effectively reduce the levels of inflammatory markers such as CRP *in vivo*. These findings validate the traditional use of plants for treating inflammatory disorders such as rheumatism and epilepsy (Van Wyk and Gericke 2018).

10.5 DERMATOLOGICAL APPLICATIONS

The increasing occurrence of drug resistance among many microorganisms is a serious concern of today's healthcare system, driving the urgent need for ongoing research to develop safer and more effective therapeutic agents. The genus *Securidaca* has been extensively studied for its potential in treating various ailments, particularly skin disorders, because its high content of phytochemicals is beneficial for skin health (Sani et al. 2023). Although research specifically addressing the dermatological applications of *Securidaca* plants is limited, the bioactive compounds found in species such as *S. longipedunculata* suggest potential benefits for skin-related treatments. The dermatological applications of *S. longipedunculata* are associated primarily with its anti-inflammatory and antioxidant properties. The antioxidant properties of the root and leaf extracts, which are attributed to their high phenolic content, can protect skin cells from oxidative stress, which is a key factor in premature aging and skin damage (Dermane et al. 2024; Muanda et al. 2010). This combination of anti-inflammatory and antioxidant properties makes *S. longipedunculata* a promising candidate for developing topical treatments for various skin disorders. Additionally, traditional uses of *Securidaca* extracts in African and Asian medicine suggest their efficacy in wound healing, possibly due to their antimicrobial and anti-inflammatory actions. However, more focused studies are needed to confirm these traditional uses and identify specific mechanisms of action in dermatological applications.

10.5.1 Wound healing and antiaging potential

Wounds are defined as disruptions in the structural and functional integrity of tissue caused by chemical, mechanical, thermal, microbiological, or immunological damage, representing a considerable global healthcare challenge owing to their substantial costs and inadequate treatment options (Goorani et al. 2019; Oluwaniyi et al. 2022). Wound healing is a multifaceted process involving a series of precisely coordinated cellular and biochemical events that restore the integrity of the skin (Hajialyani et al. 2018). Numerous therapeutic options for wound management, including analgesics, antibiotics, and nonsteroidal anti-inflammatory drugs, exist (Emmanuel et al. 2021). Medicinal herbs are a viable alternative to synthetic pharmaceuticals for wound healing because of their reduced incidence of unwanted effects. This supplemental medicine is both cost-effective and readily accessible, providing a reliable natural source of bioactive compounds with minimal or no unwanted effects (Mohamed and Osman 2017; Shedoeva et al. 2019). The skin aging process is influenced by intrinsic variables, such as genetic and hormonal changes, as well as external factors, including UV radiation, pollution, and oxidative stress (Parrado et al. 2019; Papaccio et al. 2022). The *Securidaca* genus contains several phytochemicals such as methyl salicylate and caffeic acid, which increase the degree of skin moisture supply and diminish oxidative damage, thereby alleviating signs of aging. In addition, the synergistic triad of salicylic acid, various xanthones, and esters constitute essential components of the phytochemical arsenal of this genus that enhances permeability to clear the path, neutralize inflammation to resolve cellular damage, and enhance microcirculation to stimulate tissue regeneration accordingly (Mongalo et al. 2015).

10.5.1.1 Mechanisms of wound healing

Wound healing is a complex process involving hemostasis, inflammation, proliferation, and remodeling. *Securidaca* extracts accelerate this process through multiple mechanisms (Oluwaniyi et al. 2022). As illustrated in Figure 10.2, the wound healing potential of *S. longipedunculata* is mediated

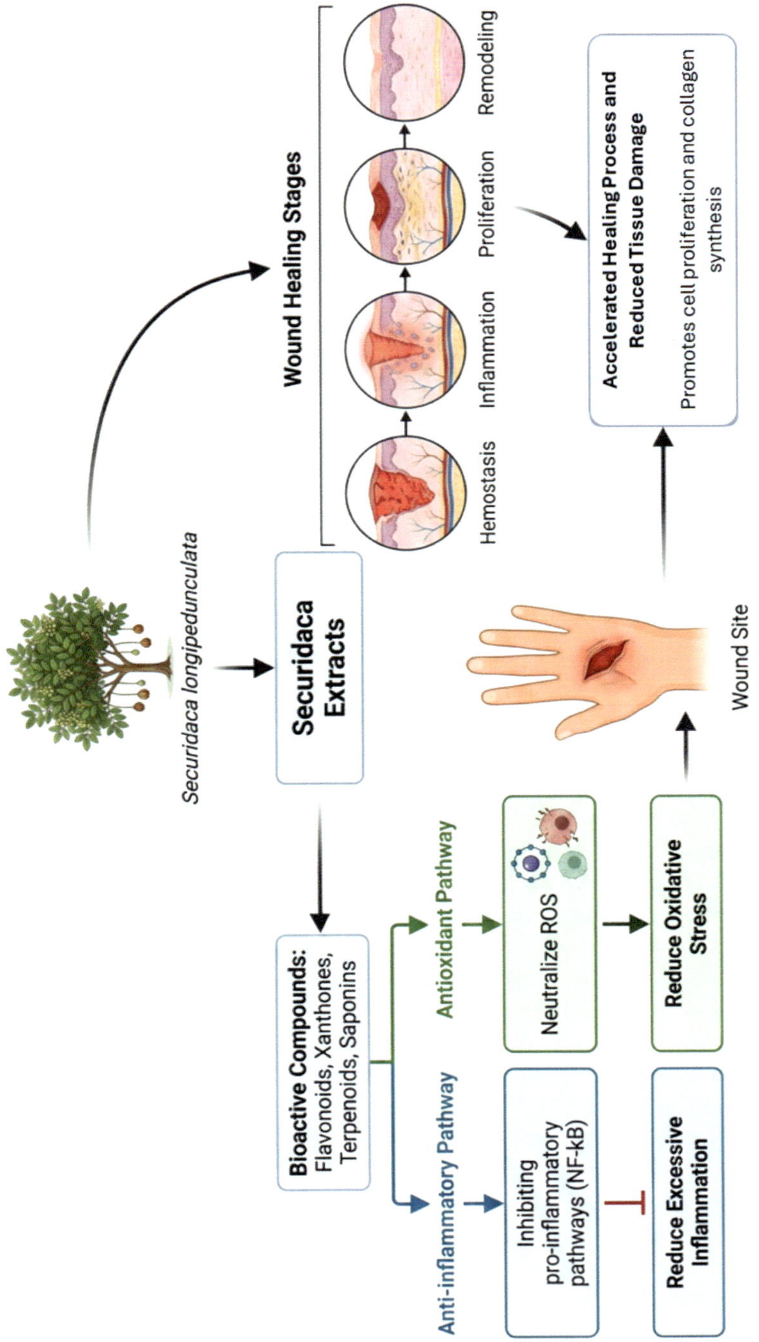

Figure 10.2 Schematic representation of the wound-healing potential of bioactive compounds from *S. longipedunculata.*

by its rich profile of bioactive compounds, including flavonoids, xanthones, terpenoids, and saponins. These phytochemicals appear to exert their therapeutic effects through a dual-mechanism approach that involves targeting the wound site.

First, the compounds engage an anti-inflammatory pathway. By inhibiting proinflammatory signaling cascades, specifically the nuclear factor kappa B (NF-κB) pathway, the extracts prevent the deleterious effects of excessive inflammation, which can delay healing (Muanda et al. 2010; Mongalo et al. 2015). Second, the compounds activate an antioxidant pathway characterized by the neutralization of reactive oxygen species (ROS), thereby significantly reducing oxidative stress at the cellular level (Namadina et al. 2020). These molecular activities support the physiological progression of the wound through its standard stages: hemostasis, inflammation, proliferation, and remodeling (Mayya et al. 2021). By balancing the inflammatory response and curbing oxidative damage, *S. longipedunculata* extracts create a favorable microenvironment that accelerates the overall healing process, limits tissue damage, and promotes crucial regenerative activities such as cell proliferation and collagen synthesis (Mongalo et al. 2015).

10.5.1.2 Antiaging and antioxidant mechanisms

Population aging is a global phenomenon, with approximately 900 million people currently aged 60 years or older worldwide. Although modern research has made significant progress in understanding the mechanisms of aging and the diseases it causes, herbal medicines are still considered among the safest and most effective treatments for aging and age-related conditions (Liu et al. 2019). Aging of the skin is driven largely by oxidative stress and the accumulation of ROS. The *Securidaca* genus combats these processes through its rich antioxidant profile. Figure 10.3 illustrates the comprehensive molecular mechanisms through which *S. longipedunculata* combats skin aging. Plant extracts rich in flavonoids, saponins, and alkaloids appear to mitigate the aging process through three primary regulatory axes (Li et al. 2024; Ungvari et al. 2018).

The extract acts as an agonist of the PI3K/Akt signaling pathway. Activation of this pathway leads to two critical downstream effects: the upregulation of telomerase, which is essential for chromosomal stability, and the inhibition of the mechanistic target of rapamycin (mTOR). The inhibition of mTOR is a key driver for inducing autophagy, a cellular "cleanup" process that removes damaged components and promotes rejuvenation (Michalak 2022). Furthermore, the activation of telomerase inhibits P21 and P53, supporting cell survival and delaying the aging process. Bioactive compounds in *S. longipedunculata* enhance antioxidant defenses by activating the Nrf2 pathway, which protects DNA from oxidative damage. Furthermore, the extract activates the Nrf2 pathway, a master regulator of antioxidant

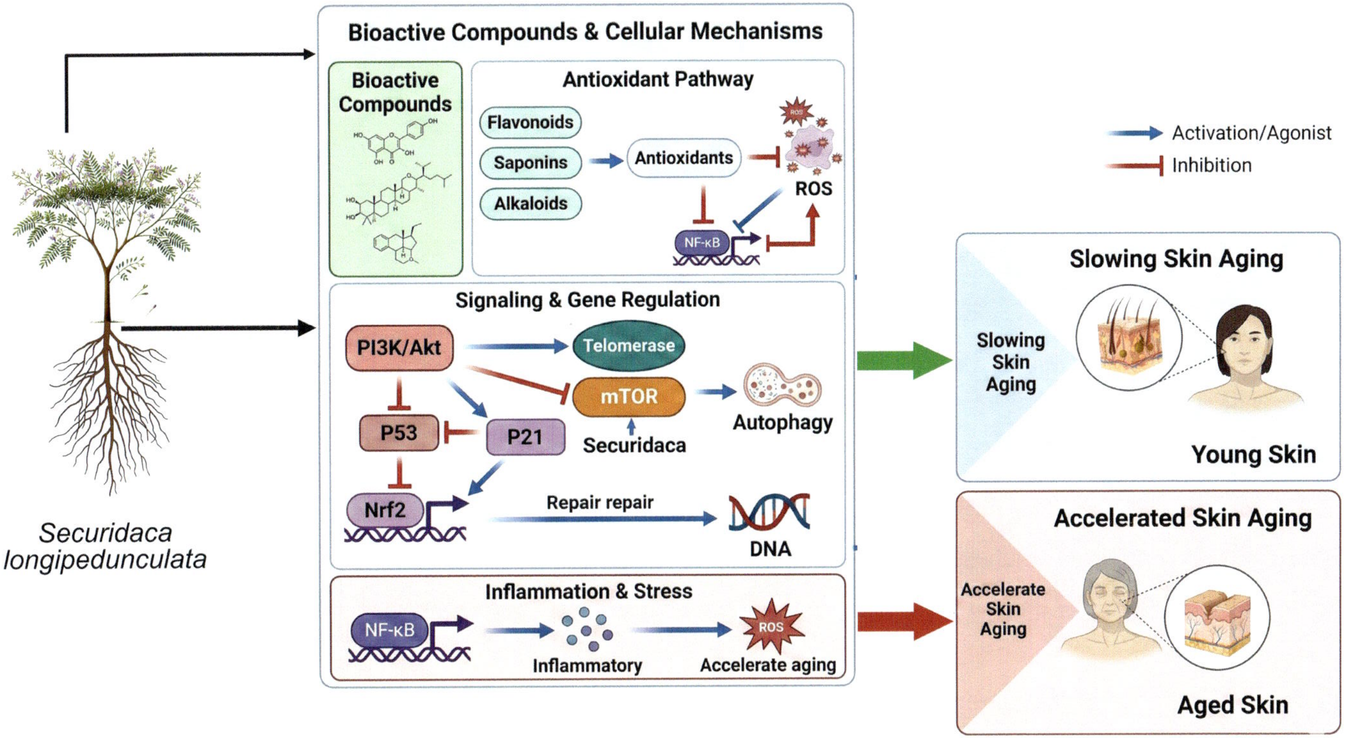

Figure 10.3 The antiaging mechanisms of *S. longipedunculata* affect the skin through bioactive compounds and cellular signaling mechanisms.

responses, thereby enhancing DNA protection against environmental stressors (Qian et al. 2023). By inhibiting the NF-κB transcription factor, the extract suppresses the inflammatory cascade. This inhibition, coupled with direct neutralization of ROS, prevents the oxidative damage that drives the accelerated aging process. The net result of these signaling modulations is the preservation of skin structural integrity and the prevention of the phenotypic characteristics of aged skin (Salminen et al. 2012; Sung et al. 2021).

10.6 PRACTICAL APPLICATIONS FOR MODERN SKIN CARE FORMULATIONS

The global interest in natural skin care solutions has increased due to the reduced side effects and better tolerance of plant-based remedies over synthetic alternatives (Mohamed and Osman 2017; Alam et al. 2022). While *Securidaca* species have a rich history in traditional medicine, their integration into modern cosmeceuticals is still in the nascent stages. Current research has focused on stabilizing bioactive extracts for topical use. For example, a hydroalcoholic extract-based ointment derived from *S. longipedunculata* root bark has demonstrated stability and local anti-inflammatory activity without causing skin irritation (Albuquerque et al. 2022). Such formulations leverage the plant's lipid-soluble compounds, such as xanthones and triterpenes, which are essential for penetrating the skin barrier. Additionally, the synthesis of biogenic silver nanoparticles using stem extracts of *S. inappendiculata* has shown promise for the development of antimicrobial skin care products with increased antioxidant capacity (Sharma et al. 2022).

With growing interest in natural skin care, incorporating medicinal plants from the *Securidaca* genus into modern formulations offers a promising opportunity for effective, sustainable, and safe solutions in skin health and wound care. These advancements pave the way for continued research into the safety and efficacy of *Securidaca*-based cosmetic formulations, supporting their use in natural skin care applications. Additionally, collaborations between researchers and cosmetic companies are facilitating clinical trials that will validate these benefits, ensuring that new products are both effective and safe for consumers (Schmidt 2012). Furthermore, the fusion of traditional herbal remedies with modern scientific techniques is likely to lead to exciting new discoveries, enhancing our understanding of how these natural ingredients can work synergistically to increase skin vitality and resilience (Dembélé et al. 2022).

10.7 CHALLENGES FOR NEW THERAPEUTIC DEVELOPMENTS

The therapeutic potential of *Securidaca* spans neuroprotection, cancer therapy, and anti-inflammatory, antidiabetic, and antimicrobial applications.

However, the development of new therapeutic applications for the *Securidaca* genus faces multiple significant challenges (Sharma et al. 2023). One of the primary obstacles is the complexity of its chemical constituents, which makes it difficult to isolate, study, and replicate its active compounds. The unique structure of compounds, such as securinine, presents challenges in synthesizing derivatives that maintain biological activity while enhancing efficacy (Klochkov and Neganova 2021). There is a lack of comprehensive studies on the active components of *Securidaca*, particularly regarding their mechanisms of action and therapeutic potential (Zuo et al. 2014).

Additionally, there is a strong need for extensive scientific research to understand the efficacy, safety, and potential side effects of these compounds thoroughly. Limited clinical trials and preclinical studies restrict the understanding of dosage, safety, and efficacy in human applications (Zuo et al. 2014). This research often requires advanced technology and significant time investment to ensure that findings are accurate and reliable. Regulatory hurdles present another barrier, as strict guidelines govern the approval of new therapeutic products. Meeting these regulatory requirements involves comprehensive clinical trials and documentation, which can be costly and time-consuming, making the path to commercialization for *Securidaca*-derived therapies challenging. The approval process for new drugs is rigorous, requiring extensive data on safety and efficacy, which can be difficult to obtain for lesser-studied plant-derived compounds (Agarwal et al. 2020). Despite these challenges, the potential of *Securidaca* compounds for treating a spectrum of skin pathologies, ranging from microbial infections to inflammatory dermatoses, suggests that further research could yield significant therapeutic advancements.

10.8 SAFETY PROFILE AND DERMAL TOXICITY

While the phytochemical potency of *Securidaca* drives its therapeutic value, it also necessitates rigorous safety considerations, particularly regarding dermal toxicity. The genus acts as a "double-edged sword," primarily due to the high concentrations of methyl salicylate in the root bark (Jayasekara et al. 2002). While beneficial as a rubefacient at low doses, in unmonitored concentrations, it acts as a severe biological irritant, capable of causing chemical burns, blistering, and necrosis of the epidermis (Mongalo et al. 2015). Furthermore, because salicylates and saponins possess high transdermal bioavailability, there is a risk of systemic absorption (salicylism) if applied to large surface areas or compromised skin. Consequently, for cosmetic applications, the use of raw, crude extracts is discouraged. Instead, safety relies on standardization and limiting the volatile methyl salicylate content with proper formulation, typically by incorporating lipid buffers (such as shea butter or carrier oils) to mitigate direct irritation while retaining the antioxidant benefits of the xanthones (Shai et al. 2024).

10.9 CONCLUSION

The *Securidaca* genus presents notable potential in dermatology, especially for wound healing and antiaging treatments, owing to its rich phytochemical composition. Key bioactive compounds such as flavonoids, alkaloids, saponins, and triterpenes provide essential dermatological benefits, including anti-inflammatory, antioxidant, and antimicrobial effects, which collectively promote skin health, enhance wound healing, and mitigate signs of aging. The long-standing traditional use of *Securidaca* across diverse cultures highlights its potential as a natural alternative in skin care, aligning with rising consumer demand for minimally processed, effective products. Integrating *Securidaca* extracts into contemporary skin care formulations could address this demand, offer sustainable solutions while minimizing the risks associated with synthetic ingredients. Future research, particularly through clinical trials, is essential to confirm the safety and efficacy of *Securidaca*-derived products, validate traditional uses, and optimize extraction and dosage for dermatological applications. Thus, *Securidaca* not only holds cultural heritage but also represents a valuable resource for innovative and sustainable skin care solutions, bridging traditional knowledge with modern dermatological science.

AUTHORS' CONTRIBUTION

Conceptualization and outline preparation: A.A.C.M. and D.H.; Data collection and curation: A.A.C.M., N.A., T.T., M.S.R., D.J.A., M.S., and D.H.; Figure preparation: A.A.C.M. and D.H.; Figures 1, 2, and 3 were created with Biorender.com; Table preparation: A.A.C.M., T.T., D.J.A., M.S.R., D.H., and N.A.; Supervision: A.A.C.M., and D.H.; Validation: A.A.C.M. and D.H.; Writing – original draft preparation: A.A.C.M, N.A., T.T., M.S.R., D.J.A., M.S., and D.H.; Writing – review and editing: A.A.C.M. K.K., and D.H. All the authors contributed toward the writing of this chapter and approved the final version submitted for publication.

CONFLICTS OF INTEREST

The authors declare that they have no conflicts of interest.

REFERENCES

Abubakar, U.S., M.S. Abdullahi, R.J. Hadiza, M. Joseph, I.K. Binta, and G.U. Habiba. 2018. Antibacterial activity of Securidaca longepedunculata stem bark against some clinical bacterial isolates. *Journal of Pharmacognosy and Phytochemistry* 7, no. 6: 1967–1970.

Adiele, R.C., B.B. Fakae, and I.U. Isuzu. 2013. Anthelmintic activity of Securidaca longepedunculata (family: Polygalaceae) root extract in mice, in vitro and in vivo. *Asian Pacific Journal of Tropical Medicine* 6, no. 11: 841–846.

Afrisham, R., M. Aberomand, M.A. Ghaffari, A. Siahpoosh, and M. Jamalan. 2015. Inhibitory effect of Heracleum persicum and Ziziphus jujuba on activity of alpha-amylase. *Journal of Botany* 2015, no. 1 (January 1): 824683. https://doi.org/10.1155/2015/824683.

Agarwal, M., D. Wald, M. Gundluru, G. Karan, and Z. Xia. 2020. Securinine and norsecurinine analogue compounds for the treatment of myeloid disorders. Google Patents.

Ahmed, L.A., B. Deka, H. Sharmah, P. Patowary, D. Bharali, B.J. Sahariah, and A. Talukdar. 2024. Exploring synthetic xanthone derivatives as potential anti-inflammatory agents: A comprehensive review. *Chemical Papers* 78, no. 13: 7313–7333.

Alam, M., S. Ahmed, A.M. Elasbali, M. Adnan, S. Alam, Md. I. Hassan, and V.R. Pasupuleti. 2022. Therapeutic implications of caffeic acid in cancer and neurological diseases. *Frontiers in Oncology* 12. www.frontiersin.org/journals/oncology/articles/10.3389/fonc.2022.860508.

Albuquerque, P.B.S., W.F. de Oliveira, P.M. dos Santos Silva, M.T. dos Santos Correia, J.F. Kennedy, and L.C.B.B. Coelho. 2022. Skincare application of medicinal plant polysaccharides – A review. *Carbohydrate Polymers* 277: 118824. www.sciencedirect.com/science/article/pii/S014486172101211X.

Al-Snafi, A.E. 2020. Oils and fats contents of medicinal plants, as natural ingredients for many therapeutic purposes – A review. *IOSR Journal of Pharmacy* 10, no. 7: 1–41.

Awoke, A., G. Gudesho, F. Akmel, and P. Shanmugasundaram. 2024. Traditionally used medicinal plants for human ailments and their threats in Guraferda District, Benchi-Sheko Zone, southwest Ethiopia. *Journal of Ethnobiology and Ethnomedicine* 20, no. 1: 82. https://doi.org/10.1186/s13002-024-00709-5.

Aymard-Corredor, G.A. 2022. Novelties in *Securidaca* (Polygalaceae, Polygaleae) for the floras of Colombia and Venezuela. *Harvard Papers in Botany* 27, no. 2 (December 1): 143–156. https://doi.org/10.3100/hpib.v27iss2.2022.n2.

Barrera-Guzmán, L.A., J. Cadena-Iñiguez, J.P. Legaria-Solano, and J. Sahagún-Castellanos. 2021. Phylogenetics of the genus Sechium P. Brown: A review. *Spanish Journal of Agricultural Research* 19, no. 1: e07R01–e07R01.

Carrol, D.H., F. Chassagne, M. Dettweiler, and C.L. Quave. 2020. Antibacterial activity of plant species used for oral health against *Porphyromonas gingivalis*. *PLOS ONE* 15, no. 10 (October 8): e0239316. https://doi.org/10.1371/journal.pone.0239316. www.sciencedirect.com/science/article/pii/S0378874120330865.

Curran, K. 2017. The ethno-herbalist review of phytochemicals. *Retrieved on August* 30: 2018.

Dalir Abdolahinia, E., S. Hajisadeghi, Z. Moayedi Banan, E. Dadgar, A. Delaramifar, S. Izadian, S. Sharifi, and S. Maleki Dizaj. 2023. Potential applications of medicinal herbs and phytochemicals in oral and dental health: Status quo and future perspectives. *Oral Diseases* 29, no. 7 (October 1): 2468–2482. https://doi.org/10.1111/odi.14276.

Datagni G. A.P. Mouzou, K. Metowogo, Y.G. Afanyibo, A. Sadji, and K. Eklu-Gadegbeku. 2021. In-vitro anti-inflammatory and antimicrobial activity of *Securidaca* longepedunculata and Annona senegalensis hydro-alcoholic extract. *Journal of Drug Delivery and Therapeutics* 11, no. 5-S: 63–70. http://doi.org/10.22270/jddt. v11i5-S.5090.

Dembélé, D.L., A. Dénou, M. Haidara, and R. Sanogo. 2022. Formulation d'une pommade antalgique et anti-inflammatoire à base d'un extrait hydroalcoolique des écorces de racines de Securidaca longepedunculata Fresen (Polygalaceae). *Journal of the Cameroon Academy of Sciences* 17, no. 3: 219–234.

Dermane, A., A.K.G. Kporvie, K.P. Kindji, K. Metowogo, and K. Eklu-Gadegbeku. 2024. Immunomodulatory and anti-inflammatory activities of hydro-ethanolic extract of Securidaca longepedunculata Fresen leaves. *Journal of Herbmed Pharmacology* 13, no. 2 (April 1): 280–288. https://herbmedpharmacol.com/Article/jhp-49352.

Desta, B. 1993. Ethiopian traditional herbal drugs. Part II: Antimicrobial activity of 63 medicinal plants. *Journal of Ethnopharmacology* 39, no. 2: 129–139. www.sciencedirect.com/science/article/pii/0378874193900284.

Emmanuel, S., A.A. Adesibikan, and O.S. Oladeji. 2021. Phytotherapeutic mechanism of medicinal plants with wound-healing potential: A mini-review. *ScienceOpen Preprints*. DOI: 10.14293/S2199-1006.1.SOR-.PPC8PLL.v1

Ezuruike, U.F. and J.M. Prieto. 2014. The use of plants in the traditional management of diabetes in Nigeria: Pharmacological and toxicological considerations. *Journal of Ethnopharmacology* 155, no. 2: 857–924. www.sciencedirect.com/science/article/pii/S0378874114004243.

Ferreira, M., A. Matos, A. Couras, J. Marto, and H. Ribeiro. 2022. Overview of cosmetic regulatory frameworks around the world. *Cosmetics* 9, no. 4: 72.

Gaobotse, G., S. Venkataraman, P.D. Brown, K. Masisi, T.E. Kwape, D.O. Nkwe, G. Rantong, and A. Makhzoum. 2023. The use of African medicinal plants in cancer management. *Frontiers in Pharmacology* 14: 1122388.

GBIF. 2024. Occurrence of *Securidaca longepedunculata* Fresen. *Global Biodiversity Information Facility (GBIF.Org)*. (Accessed date 07/10/2025)

Goorani, S., M.M. Zangeneh, M.K. Koohi, N. Seydi, A. Zangeneh, N. Souri, and M.-S. Hosseini. 2019. Assessment of antioxidant and cutaneous wound healing effects of *Falcaria vulgaris* aqueous extract in Wistar male rats. *Comparative Clinical Pathology* 28, no. 2: 435–445. https://doi.org/10.1007/s00580-018-2866-3.

Gunter, N.V., S.S. Teh, Y.M. Lim, and S.H. Mah. 2020. Natural xanthones and skin inflammatory diseases: Multitargeting mechanisms of action and potential application. *Frontiers in Pharmacology* 11: 594202.

Guo, J.-W. and S.-H. Jee. 2021. Strategies to develop a suitable formulation for inflammatory skin disease treatment. *International Journal of Molecular Sciences* 22, no. 11: 6078.

Hajialyani, M., D. Tewari, E. Sobarzo-Sánchez, S.M. Nabavi, M.H. Farzaei, and M. Abdollahi. 2018. Natural product-based nanomedicines for wound healing purposes: Therapeutic targets and drug delivery systems. *International Journal of Nanomedicine* 13, no. null (September 3): 5023–5043. www.tandfonline.com/doi/abs/10.2147/IJN.S174072.

Hamill, F.A., S. Apio, N.K. Mubiru, R. Bukenya-Ziraba, M. Mosango, O.W. Maganyi, and D.D. Soejarto. 2003. Traditional herbal drugs of southern Uganda, II: Literature analysis and antimicrobial assays. *Journal of Ethnopharmacology* 84, no. 1: 57–78. www.sciencedirect.com/science/article/pii/S0378874102002891.

Hassan, L.G., K.J. Umar, A.J. Yusuf, A.B. Rabah, and A.U. Umar. 2024. Comparative antimicrobial activity of *Securidaca longepedunculata* (Fresen) extracts on selected human pathogens. *Saudi Journal of Medical and Pharmaceutical Sciences* 10, no. 1: 26–31.

Hernandez, D.F., E.L. Cervantes, D.A. Luna-Vital, and L. Mojica. 2021. Food-derived bioactive compounds with anti-aging potential for nutricosmetic and cosmeceutical products. *Critical Reviews in Food Science and Nutrition* 61, no. 22 (December 16): 3740–3755. https://doi.org/10.1080/10408398.2020.1805407.

Imam T.S., M.M. Abdullahi, and N.T. Dabo. 2017. Comparative study on the anti-parasitic activities of Securidaca and Senna occidentalis root extracts against Trichomonas vaginalis. *International Journal of Ecotoxicology and Ecobiology* 2, no. 1: 45–51. https://doi.org/10.11648/j.ijee.20170201.16.

Iyamah, P.C. and M. Idu. 2015. Ethnomedicinal survey of plants used in the treatment of malaria in southern Nigeria. *Journal of Ethnopharmacology* 173: 287–302. www.sciencedirect.com/science/article/pii/S0378874115300246.

Jahan, F., A.A. Happy, M.M.H. Chowdhury, and M.A. Hossain. 2019. Natural herbs and spices: A great resource for skin care cosmetics. *Journal of Plant Sciences* 7: 86–99.

Jayasekara, T.K., P.C. Stevenson, S.R. Belmain, D.I. Farman, and D.R. Hall. 2002. Identification of methyl salicylate as the principal volatile component in the methanol extract of root bark of *Securidaca longepedunculata* Fres. *Journal of Mass Spectrometry* 37, no. 6 (June 1): 577–580. https://doi.org/10.1002/jms.314.

Jiang, H., C.-L. Ji, K. Yang, W. Zhang, and J. Zuo. 2018. Fatty oil from Securidaca Inappendiculata exerted therapeutic effects on adjuvant-induced arthritis in mice via suppression on fibroblast-like synoviocyte. *Kaohsiung Journal of Medical Sciences* 34, no. 11: 616–625. www.sciencedirect.com/science/article/pii/S1607551X18302626.

Johnson, C.T. 1987. Taxonomy of the African species of *Securidaca* (Polygalaceae). *South African Journal of Botany* 53, no. 1: 5–11.

Kadiri, A.B., O.M. Agboola, and F.O. Fashina. 2013. Ethnobotanical survey and phyto-anatomical studies of some common plants used for the treatment of epilepsy in some rural areas of south west Nigeria.

Karama, I., V. Ouedraogo, A. Rouamba, M. Compaoré, and M. Kiendrebeogo. 2022. Total triterpene content, antioxidant activity and acute toxicity study of *Securidaca longepedunculara* Fresen. *Journal of Drug Delivery and Therapeutics* 12, no. 2: 63–67.

Karou, S.D., T. Tchacondo, M.A.D. Tchibozo, K. Anani, L. Ouattara, J. Simpore, and C. de Souza. 2012. Screening Togolese medicinal plants for few pharmacological properties. *Pharmacognosy Research* 4, no. 2: 116–122. http://europepmc.org/abstract/MED/22518084.

Keshebo, D. and M. Choudhury. 2015. Phytochemical investigation of *Securidaca longipedunculata* (Polygalaceae) and structure elucidation of benzyl 2-hydroxy-5-methoxy benzoate. *International Journal of Current Microbiology and Applied Science* 4, no. 1: 690–695.

Kisangau, D.P., H.V.M. Lyaruu, K.M. Hosea, and C.C. Joseph. 2007. Use of traditional medicines in the management of HIV/AIDS opportunistic infections in Tanzania: A case in the Bukoba Rural District. *Journal of Ethnobiology and Ethnomedicine* 3, no. 1: 29. https://doi.org/10.1186/1746-4269-3-29.

Klochkov, S. and M. Neganova. 2021. Unique indolizidine alkaloid securinine is a promising scaffold for the development of neuroprotective and antitumor drugs. *RSC Advances* 11, no. 31: 19185–19195.

Kpodar, M.S., S.D. Karou, G. Katawa, K. Anani, H.E. Gbekley, Y. Adjrah, T. Tchacondo, K. Batawila, and J. Simpore. 2016. An ethnobotanical study of plants used to treat liver diseases in the Maritime Region of Togo. *Journal of Ethnopharmacology* 181: 263–273. www.sciencedirect.com/science/article/pii/S0378874115303019.

Kporvie, A.K.G., 2024. Pharmacological potential of Securidaca longepedunculata: A comprehensive review of traditional and scientific insights. *Asian Plant Research Journal* 12,no. 6: 115–132. DOI: 10.9734/aprj/2024/v12i6287

Lawal, R.A., M.D. Ozaslan, O.S. Odesanmi, I.D. Karagoz, I.H. Kilic, and O.A.T. Ebuehi. 2012. Cytotoxic and antiproliferative activity of Securidaca longepedunculata aqueous extract on Ehrlich ascites carcinoma cells in Swiss albino mice.

Lee, E.H., J.H. Shin, S.S. Kim, and S.R. Seo. 2021. Sinapic acid controls inflammation by suppressing NLRP3 inflammasome activation. *Cells* 10, no. 9: 2327.

Lema A., K.J. Zongo, W.L.M.E. Belem-Kabre, M. Sawadogo, H.Y. Ouoba, P.A.E.D. Sombie, N. Ouedraogo, and M.J. Bangou. 2024. Phytochemical investigation of three medicinal plants oriented against antioxidant, anti-inflammatory (*in vitro* and *in vivo*) activities and acute toxicity. *Journal of Pharmacognosy and Phytochemistry* 13, no. 5: 22–31. https://doi.org/10.22271/phyto.2024.v13.i5a.15054.

Li, J., N. Zhao, R. Xu, G. Li, H. Dong, B. Wang, Z. Li, M. Fan, and X. Wei. 2022. Deciphering the antibacterial activity and mechanism of p-coumaric acid against Alicyclobacillus acidoterrestris and its application in apple juice. *International Journal of Food Microbiology* 378: 109822. www.sciencedirect.com/science/article/pii/S016816052200294X.

Li, Y., X. Tian, J. Luo, T. Bao, S. Wang, and X. Wu. 2024. Molecular mechanisms of aging and anti-aging strategies. *Cell Communication and Signaling* 22, no. 1: 285. https://doi.org/10.1186/s12964-024-01663-1.

Liu, Y., W. Weng, R. Gao, and Y. Liu. 2019. New insights for cellular and molecular mechanisms of aging and aging-related diseases: Herbal medicine as potential therapeutic approach. *Oxidative Medicine and Cellular Longevity* 2019, no. 1 (January 1): 4598167. https://doi.org/10.1155/2019/4598167.

Lognay, G., M. Marlier, D. Seck, and É. Haubruge. 2000. The occurrence of 2-hydroxy-6-methoxybenzoic acid methyl ester in *Securidaca longepedunculata* Fresen root bark. *BASE* 4, no. 2:107–110. https://popups.uliege.be/1780-4507/index.php?id=15373.

Lou, Z., H. Wang, S. Rao, J. Sun, C. Ma, and J. Li. 2012. p-Coumaric acid kills bacteria through dual damage mechanisms. *Food Control* 25, no. 2: 550–554. www.sciencedirect.com/science/article/pii/S0956713511005044.

Mădălina-Georgiana, B., S. Imre, C. Vari, B.E. Ősz, R. Ştefănescu, A. Puşcaş, G. Jîtcă, C.-M. Matei, and A. Tero-Vescan. 2024. Assessing β-sitosterol levels in dietary supplements for benign prostatic hyperplasia: Implications for therapeutic efficacy. *Cureus* 16, no. 5: e60309. http://doi.org/10.7759/cureus.60309.

Magassouba, F.B., A. Diallo, M. Kouyaté, F. Mara, O. Mara, O. Bangoura, A. Camara, et al. 2007. Ethnobotanical survey and antibacterial activity of some plants used in guinean traditional medicine. *Journal of Ethnopharmacology* 114, no. 1: 44–53. www.sciencedirect.com/science/article/pii/S0378874107003297.

Makangara, J.J., A.M. Mshandete, E.R. Mbega, J.R. Nyika, F. Mbago, E.G. Ndilanha, R.J. Nyika, and J.J. Nyika. 2024. Review on the secondary metabolites, biological properties, and ethnomedicinal uses of the component species of the Buheri Wa Afya formula used to treat COVID-19 in Tanzania. *Phytomedicine Plus* 4, no. 1: 100508. www.sciencedirect.com/science/article/pii/S2667031323001045.

Manogaran, Y., D. Jagadeesan, K. Narain, U. Kumari, P. Anand, and S. Shanmugavelu. 2023. Antibacterial response of cinnamomum iners leaves extract and cinnamic acid derivative against pathogens that triggers periimplantitis. *Research Journal of Pharmacy and Technology* 16, no. 3: 1471–1480.

Maroyi, A. 2013. Traditional use of medicinal plants in south-central Zimbabwe: Review and perspectives. *Journal of Ethnobiology and Ethnomedicine* 9, no. 1: 31. https://doi.org/10.1186/1746-4269-9-31.

Mayya, C., S. Kharbhanda, A. Haque, and D. Bhatia. 2021. Mechanisms of collective cell migration in wound healing: Physiology and disease. In *Wound Healing Research: Current Trends and Future Directions*, ed. P. Kumar and V. Kothari, 55–74. Singapore: Springer Singapore. https://doi.org/10.1007/978-981-16-2677-7_2.

Melaku, Y., T. Getahun, M. Addisu, H. Tesso, R. Eswaramoorthy, and A. Garg. 2022. Molecular docking, antibacterial and antioxidant activities of compounds isolated from Ethiopian plants. *International Journal of Secondary Metabolite* 9, no. 2: 208–328. https://doi.org/10.21448/ijsm.1023864.

Michalak, M. 2022. Plant-derived antioxidants: Significance in skin health and the ageing process. *International Journal of Molecular Sciences* 23, no. 2: 585.

Mohamed, A.H.B. and A.A.F. Osman. 2017. Antibacterial and wound healing potential of ethanolic extract of Zingiber officinale in albino rats. *Journal of Diseases and Medicinal Plants* 3, no. 1: 1–6.

Mohan, S., S. Syam, S.I. Abdelwahab, and N. Thangavel. 2018. An anti-inflammatory molecular mechanism of action of α-mangostin, the major xanthone from the pericarp of Garcinia mangostana: An *in silico*, *in vitro* and *in vivo* approach. *Food & Function* 9, no. 7: 3860–3871.

Moiketsi, B.N., K.P.P. Makale, G. Rantong, T.O. Rahube, and A. Makhzoum. 2023. Potential of selected african medicinal plants as alternative therapeutics against multi-drug-resistant bacteria. *Biomedicines* 11, no. 10: 2605.

Mongalo, N.I., L.J. McGaw, J.F. Finnie, and J. Van Staden. 2015. *Securidaca longepedunculata* Fresen (Polygalaceae): A review of its ethnomedicinal uses, phytochemistry, pharmacological properties and toxicology. *Journal of Ethnopharmacology* 165: 215–226. www.sciencedirect.com/science/article/pii/S0378874115001208.

Motlhanka, D.M.T. and G.P. Nthoiwa. 2012. Ethnobotanical survey of medicinal plants of Tswapong North, in eastern Botswana: A case of plants from Mosweu and Seolwane villages. *European Journal of Medicinal Plants* 3, no. 1: 10–24.

Muanda, F.N., A. Dicko, and R. Soulimani. 2010. Assessment of polyphenolic compounds, *in vitro* antioxidant and anti-inflammation properties of *Securidaca longepedunculata* root barks. *Comptes Rendus. Biologies* 333, no. 9: 663–669.

Musa, A.A., A.O. Oyewale, I.G. Ndukwe, S.E. Yakubu, and M.S. Abdullahi. 2013. Phytochemical screening and antimicrobial activity of solvent fractions of *Securidaca longepedunculata* (Fresen) root bark methanol extract. *Journal of Chemical and Pharmaceutical Research* 5, no. 10: 28–33.

Mustapha, A.A. 2013. Ethno-medico-botanical uses of Securidaca longepedunculata (Fresen) (family Polygalaceae) from Keffi Local Government, Nasarawa State, Nigeria. *Journal of Natural Remedies* 13: 133–137.

Nair P.P., V. Krishnakumar, and P.G. Nair. 2024. Chronic Inflammation: Cross linking insights from Ayurvedic Sciences, a silver lining to systems biology and personalized medicine. *Journal of Ayurveda and Integrative Medicine* 15, no. 4 (Jul-Aug): 101016. https://doi.org/10.1016/j.jaim.2024.101016. Epub 2024 Jul 16. PMID: 39018639; PMCID: PMC11298630.

Namadina, M.M., R.S. Shawai, F.M. Musa, U. Sunusi, M.A. Aminu, Y. Nuhu, and A.M. Umar. 2020. Phytochemical and antimicrobial activity of *Securidaca longipedunculata* root against urinary tract infection pathogens. *ChemSearch Journal* 11, no. 2: 90–98.

Ndamitso, M.M, J.O. Tijani, and S. Idris. 2013. Phytochemical and antibacterial activity of *Securidaca longepedunculata* on selected pathogens. *African Journal of Microbiology Research* 7, no. 50: 5652–5656.

Ngulde, S.I., U.K. Sandabe, R. Abounader, T.K. Dawson, Y. Zhang, I. Iliya, and I.M. Hussaini. 2019. Ethanol extract of *Securidaca longipedunculata* induces apoptosis in brain tumor (U87) cells. *BioMed Research International* 2019, no. 1 (January 1): 9826590. https://doi.org/10.1155/2019/9826590.

Nguyen, L.T., A.C. Fărcaş, S.A. Socaci, M. Tofană, Z.M. Diaconeasa, O.L. Pop, and L.C. Salanță. 2020. An overview of saponins—A bioactive group. *Bulletin of University of Agricultural Sciences and Veterinary Medicine Cluj-Napoca. Food Science and Technology* 77, no. 1: 25–36. https://doi.org/10.15835/buasvmcn-fst:2019.0036

Nordeng, H., W. Al-Zayadi, D. Diallo, N. Ballo, and B.S. Paulsen. 2013. Traditional medicine practitioners' knowledge and views on treatment of pregnant women in three regions of Mali. *Journal of Ethnobiology and Ethnomedicine* 9, no. 1: 67. https://doi.org/10.1186/1746-4269-9-67.

Obasi, T.C., C. Braicu, B.C. Iacob, E. Bodoki, A. Jurj, L. Raduly, I. Oniga, I. Berindan-Neagoe, and R. Oprean. 2018. *Securidaca*—saponins are natural inhibitors of AKT, MCL-1, and BCL2L1 in cervical cancer cells. *Cancer Management and Research* 10, no. null (November 15): 5709–5724. www.tandfonline.com/doi/abs/10.2147/CMAR.S163328.

Okoli, C.O., Akah, P.A., Ezugworie, U., 2005. Anti-inflammatory activity of extracts of root bark of *Securidaca longepedunculata* Fres (Polygalaceae). *African Journal of Traditional Complementary and Alternative Medicine* 2, 54–63.

Oluwaniyi, O.O., A.A. Adesibikan, and S.S. Emmanuel. 2022. Evaluation of wound-healing activity of *Securidaca longepedunculata* root extract in male Wistar rats. *ChemistrySelect* 7, no. 26 (July 14): e202200711. https://doi.org/10.1002/slct.202200711.

Ouedraogo, L., J. Endl, P.A.E.D. Sombié, H. Schaefer, and M. Kiendrebeogo. 2020. Ethnobotanical use and conservation assessment of medicinal plants sold in markets of Burkina Faso. *Ethnobotany Research and Applications* 20, no. 0 (October 28): 1–25. https://ethnobotanyjournal.org/index.php/era/article/view/2259.

Pandita, D. and A. Pandita. 2021. Secondary metabolites in medicinal and aromatic plants (MAPs): Potent molecules in nature's arsenal to fight human diseases. In *Medicinal and Aromatic Plants: Healthcare and Industrial Applications*, ed. T. Aftab and K.R. Hakeem, 41–84. Cham: Springer International Publishing. https://doi.org/10.1007/978-3-030-58975-2_2.

Pang, X., E.A. Makinde, F.N. Eze, and O.J. Olatunji. 2021. *Securidaca inappendiculata* polyphenol rich extract counteracts cognitive deficits, neuropathy, neuroinflammation and oxidative stress in diabetic encephalopathic rats via p38 MAPK/Nrf2/HO-1 pathways. *Frontiers in Pharmacology* 12: 737764. www.frontiersin.org/journals/pharmacology/articles/10.3389/fphar.2021.737764.

Papaccio, F., S. Caputo, and B. Bellei. 2022. Focus on the contribution of oxidative stress in skin aging. *Antioxidants* 11, no. 6: 1121.

Parrado, C., S. Mercado-Saenz, A. Perez-Davo, Y. Gilaberte, S. Gonzalez, and A. Juarranz. 2019. Environmental stressors on skin aging: Mechanistic insights. *Frontiers in Pharmacology* 10: 759.www.frontiersin.org/journals/pharmacology/articles/10.3389/fphar.2019.00759.

Poudiougou, O., P.-E. Bruand, P.P. Mounkoro, J.-M. Gaglione, K. Nimaga, M. Sy, C. Vincent, F. Calas, A. Fall-Ndao, and L. Petiteau. 2021. Mental health capacity building in Mali by training rural general practitioners and raising community awareness. *Pan African Medical Journal* 38, no. 1: 389. www.ajol.info/index.php/pamj/article/view/241246

Qian, H. Y. Shan, R. Gong, D. Lin, M. Zhang, C. Wang, and L. Wang. 2023. Mechanism of action and therapeutic effects of oxidative stress and stem cell-based materials in skin aging: Current evidence and future perspectives. *Frontiers in Bioengineering and Biotechnology* 10: 1082403.www.frontiersin.org/journals/bioengineering-and-biotechnology/articles/10.3389/fbioe.2022.1082403.

Qian, L.-S. J.-H. Chen, T. Deng, and H. Sun. 2020. Plant diversity in Yunnan: Current status and future directions. *Plant Diversity* 42, no. 4: 281–291. www.sciencedirect.com/science/article/pii/S2468265920300676.

Salminen, A., A. Kauppinen, and K. Kaarniranta. 2012. Phytochemicals suppress nuclear factor-κB signaling: Impact on health span and the aging process. *Current Opinion in Clinical Nutrition & Metabolic Care* 15, no. 1: 22–28.https://journals.lww.com/co-clinicalnutrition/fulltext/2012/01000/phytochemicals_suppress_nuclear_factor__b.5.aspx.

Sani, A., N.A. Ashezi, U.S. Abubakar, S. Garba, Z.A. AbdulAziz, M.R. Bukar, and M.T. Sulaiman. 2023. Phytochemical screening and antimicrobial activity of methanolic leaf extract of *Securidaca longipedunculata* Fresen (Polygalaceae). *Journal of Current Biomedical Research* 3, no. 2: 921–932.

Santos-Buelga, C., A.M. González-Paramás, and S. González-Manzano. 2023. Phenolic acids and derivatives: Description, sources, properties, and applications. In *Natural Secondary Metabolites: From Nature, Through Science, to Industry*, ed. M. Carocho, S.A. Heleno, and L. Barros, 37–72. Cham: Springer International Publishing. https://doi.org/10.1007/978-3-031-18587-8_3.

Saraiva, M.E., A.V.R. de A. Ulisses, D.A. Ribeiro, L.G.S. de Oliveira, D.G. de Macêdo, F. de F.S. de Sousa, I.R.A. de Menezes, E.V. de S.B. Sampaio, and M.M. de A. Souza. 2015. Plant species as a therapeutic resource in areas of the savanna in the state of Pernambuco, northeast Brazil. *Journal of Ethnopharmacology* 171: 141–153. www.sciencedirect.com/science/article/pii/S0378874115003669.

Schmidt, B.M. 2012. Responsible use of medicinal plants for cosmetics. *HortScience Horts* 47, no. 8: 985–991. https://journals.ashs.org/hortsci/view/journals/hortsci/47/8/article-p985.xml.

Semenya, S.S. and M.J. Potgieter. 2013. Ethnobotanical survey of medicinal plants used by Bapedi traditional healers to treat erectile dysfunction in the Limpopo province, south africa. *Journal of Medicinal Plants Research* 7, no. 7: 349–357.

Shai, K., Z.B. Hassan, S. Lebelo, J.W. Ng'ambi, M. Mabelebele, and N.A. Sebola. 2024. Chemical analysis and biological activities of various parts of *Securidaca longipedunculata* from South Africa. *Natural Product Communications* 19, no. 9 (September 1): 1934578X241261018. https://doi.org/10.1177/1934578X241261018.

Sharma, K., R. Kaur, S. Kumar, R.K. Saini, S. Sharma, S. V Pawde, and V. Kumar. 2023. Saponins: A concise review on food related aspects, applications and health. *Food Chemistry Advances* 2: 100191. www.sciencedirect.com/science/article/pii/S2772753X23000114.

Sharma, M., R. Chadha, and N. Dhingra. 2017. Phytotherapeutic agents for benign prostatic hyperplasia: An overview. *Mini Reviews in Medicinal Chemistry* 17, no. 14: 1346–1363.

Sharma, R., R.A. Deep, and S.T. Abdullah. 2022. Herbal products as skincare therapeutic agents against ultraviolet radiation-induced skin disorders. *Journal of Ayurveda and Integrative Medicine* 13, no. 1: 100500. www.sciencedirect.com/science/article/pii/S0975947621001480.

Shedoeva, A., D. Leavesley, Z. Upton, and C. Fan. 2019. Wound healing and the use of medicinal plants. *Evidence-Based Complementary and Alternative Medicine* 2019, no. 1 (January 1): 2684108. https://doi.org/10.1155/2019/2684108.

Shemishere, U.B., D.A. Anyebe, A.Y. Bashir, J. Emmanuel, J. Ifie, and T. Yahaya. 2020. Phytochemical screening and free radical scavenging activities of methanol leaf and flower extract of *Securidaca longipedunculata*. *Fudma Journal of Sciences* 4, no. 1 (April 13): 37–42. https://fjs.fudutsinma.edu.ng/index.php/fjs/article/view/15.

Suleiman Abubakar, U., U.H. Danmalam, H. Ibrahim, B.B. Maiha, H. Rasheed-Jada, and M. Sani Abdullahi. 2022. Phytochemical investigation of the stem bark of *Securidaca longipedunculata* Fresen (Polygalaceae) (I). *Journal of Current Biomedical Research* 2, no. 1 (February 28): 38–49. https://journals.unizik.edu.ng/jcbr/article/view/880.

Sung, J.Y., S.G. Kim, J.-R. Kim, and H.C. Choi. 2021. SIRT1 suppresses cellular senescence and inflammatory cytokine release in human dermal fibroblasts by promoting the deacetylation of NF-κB and activating autophagy. *Experimental Gerontology* 150: 111394. www.sciencedirect.com/science/article/pii/S0531556521001765.

Tabuti, J.R.S. C.B. Kukunda, D. Kaweesi, and O.M.J. Kasilo. 2012. Herbal medicine use in the districts of Nakapiripirit, Pallisa, Kanungu, and Mukono in Uganda. *Journal of Ethnobiology and Ethnomedicine* 8, no. 1: 35. https://doi.org/10.1186/1746-4269-8-35.

Tian, X.R. H.F. Tang, X.L. Tian, J.J. Hu, L.L. Huang, and K.R. Gustafson. 2018. Review of bioactive secondary metabolites from marine bryozoans in the progress of new drugs discovery. *Future Medicinal Chemistry* 10, no. 12 (June 1): 1497–1514. https://doi.org/10.4155/fmc-2018-0012.

Tikisa, T., D. Abdissa, and N. Abdissa. 2019. Chemical constituents of *Securidaca longipedunculata* root bark and evaluation of their antibacterial activities. *Ethiopian Journal of Education and Sciences* 14, no. 2: 1–8.

Toudji, G.A. E. Thiombiano, S.D. Karou, K. Anani, Y. Adjrah, H.E. Gbekley, M. Kiendrebeogo, Y. Ameyapoh, and J. Simpore. 2018. Antibacterial and anti-inflammatory activities of crude extracts of three Togolese medicinal plants against ESBL Klebsiella pneumoniae strains. *African Journal of Traditional, Complementary and Alternative Medicines* 15, no. 1: 42–58.

Ungvari, Z., S. Tarantini, A.J. Donato, V. Galvan, and A. Csiszar. 2018. Mechanisms of vascular aging. *Circulation Research* 123, no. 7 (September 14): 849–867. https://doi.org/10.1161/CIRCRESAHA.118.311378.

Valentin, B.C., O.N. Philippe, M.M. Henry, B.A. Salvius, M.K. Suzanne, F.M. Kasali, and L.S.J. Baptiste. 2024. Ethnomedical knowledge of plants used in nonconventional medicine for wound healing in Lubumbashi, Haut-Katanga Province, DR Congo. *The Scientific World Journal* 2024, no. 1 (January 1): 4049263. https://doi.org/10.1155/2024/4049263.

Van Wyk, B.E. and N. Gericke. 2018. *People's Plants: A Guide to Useful Plants of Southern Africa* (second ed.). Pretoria: Briza Publications.

Wang, L., F. Pan, and T. Luo. 2021. Sinapic acid attenuates rheumatoid arthritis through reducing inflammation and oxidative stress by downregulating IκB kinase. *Journal of Interferon & Cytokine Research* 41, no. 9 (September 1): 347–354. https://doi.org/10.1089/jir.2021.0044.

Wang, Q., C. Ma, Y. Ma, X. Li, Y. Chen, and J. Chen. 2017. Structure–activity relationships of diverse xanthones against multidrug resistant human tumor cells. *Bioorganic & Medicinal Chemistry Letters* 27, no. 3: 447–449. www.sciencedirect.com/science/article/pii/S0960894X16313129.

Wanzala, W., S.M. Syombua, and J.O. Alwala. 2016. A survey of the applications and use of ethnomedicinal plants and plant products for healthcare from the Ukambani region in Eastern Kenya. *Indian Journal of Ethnophytopharmaceuticals* 2, no. 2: 6–58. http://hdl.handle.net/123456789/9590

Wong, K.W., S. S. Teh, K.P. Law, I.S. Ismail, K. Sato, N. Mase, and S. H. Mah. 2023. Synthesis of benzylated amine-substituted xanthone derivatives and their antioxidant and anti-inflammatory activities. *Archiv Der Pharmazie* 356, no. 1 (January 1): 2200418. https://doi.org/10.1002/ardp.202200418.

Wu, Y. J., Bai, K. Zhong, Y. Huang, H. Qi, Y. Jiang, and H. Gao. 2016. Antibacterial activity and membrane-disruptive mechanism of 3-p-trans-coumaroyl-2-hydroxyquinic acid, a novel phenolic compound from pine needles of Cedrus deodara, against *Staphylococcus aureus*. *Molecules* 21, no. 8: 1084.

Zulu, D., B.L.K. Thokozani, G.W. Sileshi, Z. Teklehaimanot, D.S.B. Gondwe, V. Sarasan, and P.C. Stevenson. 2011. Propagation of the African medicinal and pesticidal plant, *Securidaca longepedunculata*. *African Journal of Biotechnology* 10, no. 32: 5988–5992.

Zuo, J., Y. Xia, X. Li, and J. Chen. 2014. Xanthones from *Securidaca inappendiculata* exert significant therapeutic efficacy on adjuvant-induced arthritis in mice. *Inflammation* 37, no. 3: 908–916. https://doi.org/10.1007/s10753-014-9810-8.

Index

Note: Figures are indicated by *italics*. Tables are indicated by **bold**.